HIGH RESOLUTION NMR IN SOLIDS

SELECTIVE AVERAGING

Advances in
MAGNETIC RESONANCE

EDITED BY

JOHN S. WAUGH

DEPARTMENT OF CHEMISTRY
MASSACHUSETTS INSTITUTE OF TECHNOLOGY
CAMBRIDGE, MASSACHUSETTS

HIGH RESOLUTION NMR IN SOLIDS

SELECTIVE AVERAGING

SUPPLEMENT 1
Advances in Magnetic Resonance

ULRICH HAEBERLEN
MAX-PLANCK-INSTITUT FÜR MEDIZINISCHE FORSCHUNG
HEIDELBERG, GERMANY

1976

ACADEMIC PRESS   New York San Francisco London

A Subsidiary of Harcourt Brace Jovanovich, Publishers

COPYRIGHT © 1976, BY ACADEMIC PRESS, INC.
ALL RIGHTS RESERVED.
NO PART OF THIS PUBLICATION MAY BE REPRODUCED OR
TRANSMITTED IN ANY FORM OR BY ANY MEANS, ELECTRONIC
OR MECHANICAL, INCLUDING PHOTOCOPY, RECORDING, OR ANY
INFORMATION STORAGE AND RETRIEVAL SYSTEM, WITHOUT
PERMISSION IN WRITING FROM THE PUBLISHER.

ACADEMIC PRESS, INC.
111 Fifth Avenue, New York, New York 10003

United Kingdom Edition published by
ACADEMIC PRESS, INC. (LONDON) LTD.
24/28 Oval Road, London NW1

LIBRARY OF CONGRESS CATALOG CARD NUMBER: 65-26774

ISBN 0-12-025561-8

PRINTED IN THE UNITED STATES OF AMERICA

narrow is beautiful

Contents

Foreword

Over the past several years a number of sophisticated new experimental methods have emerged which considerably broaden the scope of NMR, particularly, but not exclusively, for the study of solids and semisolids. Among these one of the first, and in many ways the most interesting, has been the family of techniques that has come to be called "multiple-pulse NMR." Followers of this subject will be aware that Ulrich Haeberlen took part in this development from its beginning stages and has continued to be perhaps its most consistent and successful exponent.

Although multiple-pulse NMR has been around for several years and has aroused a certain amount of interest, there has been no self-contained exposition of this subject from which an outsider could learn about its theory, practice, and domain of applicability. It was with that in mind that I asked Dr. Haeberlen to contribute a chapter on the subject to *Advances in Magnetic Resonance*. It is a measure of the growth of the field that his contribution outgrew the confines of a single chapter and is being published as a supplementary volume to this serial publication.

No more authoritative person could have been found to write a monograph on multiple pulse and NMR, and I believe that this volume represents the last word on the subject.

JOHN S. WAUGH

Preface

"High resolution NMR in solids" has become the standard term for a variety of experimental techniques which allow the observation in solids of resolved, "chemically shifted" NMR lines from spin-$\frac{1}{2}$ nuclei in magnetically equivalent lattice positions. This is usually accomplished at a level of spectral resolution which may seem ridiculously poor by the standards of high resolution NMR proper. But this shortcoming of high resolution NMR in solids is compensated for by "new" information accessible to the experimenter; the most important one concerns nuclear magnetic shielding tensors. A substantial part of this volume is consequently devoted to the determination of nuclear magnetic shielding tensors (Chapters III and VI).

Four main approaches to high resolution NMR in solids are currently in use: *magic-angle sample-spinning, multiple-pulse, proton-enhanced nuclear induction,* and *indirect detection* methods. They have been tailored to suit a variety of different experimental situations. In this volume we discuss the principles of how "high resolution" is achieved for all of them (Chapter IV). The latter two also involve, as well as means for obtaining "high resolution," ingenious tricks for enhancing the *sensitivity* of detecting the NMR signal; but we consider these as being outside the scope of this volume.

Experimental and theoretical details as well as a comprehensive review of applications are restricted to *multiple-pulse techniques* (Chapters V and VI). It is in this field that I have concentrated my own research efforts in recent years.

The *leitmotiv* of high resolution NMR—*generally,* not only in solids— is always some kind of *selective averaging.* By this we mean that those interactions of the nuclear spins which are considered uninteresting in a particular experiment are somehow made time dependent. Ways to accomplish this range from simply melting the sample to applying highly complex multiple-pulse sequences. Provided the time dependences so introduced meet certain conditions, the *unwanted* spin interactions are efficiently averaged out, whereas the interesting ones remain more or less unaffected.

In order to understand what *is* possible and what is *not* by selective averaging, a good grasp of the tensorial properties of nuclear spin interactions in both ordinary and spin spaces is required. Chapter II is devoted to a study of these properties.

Chapter III deals with the manifestations of nuclear magnetic shielding in NMR spectra of both single-crystal and powder samples. The techniques for analyzing spectra and "rotation patterns" in terms of shielding tensors are discussed. Line broadening by spin–spin interactions is largely disregarded in this chapter.

Chapter IV is the central one. Here we treat a wide range of phenomena in NMR which are the result of intentional or natural, selective or unselective averaging processes. Some can be described adequately by simple "average Hamiltonians," others require the inclusion of corrections. A theory (average Hamiltonian theory) that yields these corrections in a general way is outlined in Chapter IV, Section D.

Chapter V is a detailed discussion of multiple-pulse sequences intended for high resolution NMR in solids. It is rather specialized and mainly written for readers interested in the development and the limitations of the technique.

Chapter VI is a review of applications of multiple-pulse techniques. The emphasis is—naturally—on measurements of ^{19}F and ^{1}H shielding tensors. I hope I have not overlooked important contributions. As it is likely that I have, I want to apologize "preventively" to the authors. The current status of the interpretation of shielding tensors is also discussed. The viewpoint and the language are naturally those of an experimentalist.

What is lacking—fully by intention—is a chapter on instrumentation, although Chapter V and Appendix D do contain technical hints here and there. Multiple-pulse spectrometers can be assembled to a large extent from commercially available components.

From talking to several people interested in multiple-pulse techniques I feel, though, that one word of warning is in place: The crucial part of a multiple-pulse spectrometer is *not* the pulse programmer. From the point of view of being difficult and critical in design, construction, and operation, I would even call it a *quantité négligeable*. In my opinion the crucial parts are the transmitter and the probe. Readers interested in instrumentation are recommended to consult a recent article by the MIT NMR group in Volume 5 of this serial publication.

As regards notation I have always tried to keep it self-explanatory, even at the expense of uniformity. Hamiltonians are written both in dimensions of energy ($\hbar\mathcal{H}$) and angular velocity ($\mathcal{H}$). Unfortunately, I have had to decorate the symbol for the magnetogyric ratio, γ_n, with an index n in order to distinguish it from the frequently occurring Euler angle γ.

It is a pleasure for me to take this opportunity to express my gratitude to Professor J. S. Waugh for his suggestion that I review multiple-pulse NMR for *Advances in Magnetic Resonance*, for his kind encouragement during the writing of what became this volume, and—last but not least—for the exciting and stimulating couple of years I could work in his laboratory. To Professor K. H. Hausser I am greatly indebted for his continuous interest in the progress of this work and, above all, for enabling me and my close colleagues to set up and run an excellently equipped laboratory devoted to high resolution NMR in solids. It is probably only natural that I draw heavily on the work and ideas of many of my colleagues. I acknowledge gratefully their direct and indirect contributions to this work. My special thanks are due to my friend H. W. Spiess for innumerable discussions from which this work profited a good deal and specifically for critically reading the manuscript. The beautiful dipolar and multiple-pulse spectra of KHF_2 I owe to P. Van Hecke who spent the first half of this year (1975) in our laboratory. I want to thank P. Moravek for drawing most of the figures and diagrams. My wife painstakingly typed a first draft and the final version of the manuscript. I am very grateful to her for this help, but even more so for the good humor with which she endured my virtually total preoccupation with the preparation of the manuscript for more than a full year.

ULRICH HAEBERLEN

I. Introduction

NMR shares with all other branches of spectroscopy a perpetual desire to improve the resolution of its spectra. Better resolved spectra usually imply more information about the sample or object under investigation. One part of the hostile line broadening is always due to the apparatus (although occasionally it may be totally negligible), while another arises from the physical properties of the sample at hand. Fighting the former is an engineering problem. In the pioneer years of high-resolution NMR in liquids it was, indeed, a very severe problem. Nowadays for NMR in liquids it can be considered to have been solved for most practical purposes. Nevertheless, the instrumental limitation of resolution is still one of the most important specifications of a high-resolution NMR apparatus.

The situation is quite different for NMR in solids. Even in the early days of NMR spectroscopy in solids the instrumental contribution to the linewidths was usually very small in comparison to the physically inherent widths of the lines. The major source of NMR line broadening in solids is the dipolar coupling of the nuclear spins. The nuclear spins "see" not only the applied field, but in addition dipolar fields originating in the nuclear magnetic moments associated with the spins of all neighbor nuclei. It is the magnitude of the vector sum of all magnetic fields at the site of a given nucleus that determines where that nucleus will "appear" in the NMR spectrum. As the applied field usually exceeds by several orders of magnitude the internal dipolar fields, only those components of the latter matter that are parallel or antiparallel to the applied field. This statement contains what is usually called "truncation" or restriction to "secular" terms of the dipolar Hamiltonian. We shall come back to this topic in detail in Chapter IV.

Whether the field contribution from a neighbor nucleus is parallel or antiparallel to the applied field depends among other things upon the spin orientation of that nucleus. Under normal circumstances the spins are oriented almost completely at random. The dipolar field that a nuclear spin "sees" is

1

therefore the sum of contributions the signs of which are unrelated and equally likely to be plus or minus.

Therefore, for the ensemble of spins in a macroscopic sample there exists a distribution of the magnitudes of the magnetic field, which is reflected in the NMR spectrum of the sample as the so-called dipolar line broadening. For samples containing abundant spins with "large" magnetogyric ratios, such as protons or fluorines, a typical value of the root mean square width of the dipolar field distribution is 10 G. If the same nuclei are observed by NMR, the lines will have a width of about 40 kHz.

By contrast, NMR lines in liquids are typically as narrow as 0.1–1 Hz. Why are they so narrow? After all, by melting a solid with an NMR linewidth of 40 kHz the nuclear spins and their magnetic moments do not just disappear. As is probably familiar to anyone who undertakes to read this volume the reason for the narrow NMR lines of liquid samples is the rapid, random, isotropic reorientational and translational motions of molecules in liquids. The components of the dipolar fields parallel to the applied field that the nuclear spins present to each other are not only dependent upon the nuclear spin orientations but also upon the orientations of their internuclear vectors with respect to the applied field. It is via this spatial orientational dependence of the dipolar fields that molecular motions are able to modulate them.

Provided the time scale of the molecular motions, and hence of the modulation of the dipolar fields, is considerably shorter than the inverse of the dipolar linewidth of the solid sample any reference spin "sees" effectively only the time average of the (secular) dipolar fields of its neighbors. This phenomenon of "motional averaging" is by no means restricted to dipolar fields, nor to NMR, but is completely general and very widespread in physics. We shall treat it mathematically in Chapter IV.

The time scale of molecular reorientations in nonviscous liquids is, on the one hand, 10^{-10}–10^{-11} sec—which is much shorter than the inverse dipolar linewidths of the corresponding solids—and, on the other hand, the average of the secular dipolar fields over isotropic molecular motions is zero. This explains why NMR lines of liquids (and gases) are so narrow.

The trouble with obtaining high resolution by random, isotropic molecular reorientational motions is that not only the dipolar fields of the nuclei, but all anisotropic interactions of the spins are averaged out at the same time. That means the positions of the NMR lines in spectra of liquids—where we can measure them with high precision—carry no information about any kind of anisotropic interaction of the spins, for instance, anisotropic nuclear magnetic shielding. Spectral information about anisotropic spin interactions can only be obtained from NMR spectra of solids—where the lines are broad and where, consequently, the line positions often cannot be measured accurately enough to extract the desired information.

In order to get access experimentally to this kind of information a number of ingenious schemes have been worked out. They all rely on "motional averaging" as described above, but abandon random, isotropic, fast molecular motions as the means for introducing the "motion" into the system. To achieve "selective averaging" more specific, anisotropic motions are used instead. We shall consider a variety of them in Chapter IV.

Not all of them rely upon motions in ordinary coordinate space. Recall that the (secular) dipolar field set up at the site of some reference spin r by its neighbors (denoted i) depends on both the lengths and orientations of the internuclear vectors $\mathbf{r}_{ir}$ and on the orientations of the spins i. Analogous statements apply, of course, to other types of nuclear spin–spin interactions. Molecular motions affect the dipolar fields via the latter's spatial orientational dependence. They can, obviously, also be affected via their spin orientational dependence.

This has been exploited for many years for a number of purposes including resolution enhancement. Traditionally the techniques used are not described as "averaging" but as "spin-decoupling."

We want to emphasize, however, that all spin-decoupling techniques used in NMR can be understood in a way completely analogous to "motional averaging in spatial coordinate space" as "motional averaging in nuclear spin spaces."

In the same way nature provides motions in spatial coordinate space— thermal molecular motions—it also provides motions in spin spaces. These thermal spin fluctuations assure that nonequilibrium distributions of spin orientations disappear in the course of time.

With respect to averaging of nuclear spin–spin interactions there is one important quantitative difference between thermal spin fluctuations and thermal molecular reorientations: the latter are usually fast, at least in nonviscous fluids in comparison with nuclear spin—spin interactions, whereas the former are fast only in exceptional cases. In Chapter VI we shall cite one such case (proton-iodine "self-decoupling" in *trans*-diiodoethylene).

In order to achieve resolution enhancement by motional averaging in spin spaces the nuclear spins must, therefore, usually be stirred artificially. This can be done by irradiating the spins with properly chosen rf fields. High-resolution NMR spectroscopists and makers of high-resolution NMR equipment distinguish hetero- from homodecoupling. By heterodecoupling they mean stirring spins belonging to a spin species other than the observed spins by applying an rf field at the Larmor frequency of the former spin species. They speak of homodecoupling when the irradiated and the observed spins belong to the same spin species. In this case, however, the observed and the irradiated spins must be "chemically shifted" so that the difference between homo- and heterodecoupling really is only a quantitative or technical one, and not one of principle. The need to differentiate between stirred and observed spins or more generally between stirred and observed degrees of freedom seems to erect a barrier that prevents the experimenter from over-coming the dipolar coupling of *like* spins in solids. This barrier, however, exists only as long as one takes into consideration only random motions. Stirring the observed spins violently in a random manner means inevitably losing the desired spectral information.

However, the barrier can be overcome by bringing into play coherent motions. Coherent motions can be imposed on the spatial degrees of freedom of the system by physically rotating ("spinning") the sample. Depending on the orientation of the axis of rotation with respect to the applied field a variety of line-broadening spin interactions, including dipolar couplings between like spins, can be suppressed in this way.

Coherent motions can be imposed directly on the observed spins by exposing them to coherent, either cw or pulsed rf, irradiation. Because of the lack of inertia effects it is easier in practice to impose coherent motions on spin than on spatial degrees of freedom. The price one has to pay is that one must follow certain rules in observing the nuclear spins (see Chapter IV).

While sample spinning about an axis perpendicular to the applied field has been crucial for the development of high-resolution NMR in liquids it was the "invention" of the so-called magic-angle sample-spinning technique that marked the beginning of high-resolution NMR in solids; but it was only the introduction of coherent averaging techniques in spin spaces that really gave it the final boost it needed to become firmly established.

Coherent motions in either spin or coordinate spaces do not "just enhance generally" the resolution of NMR spectra. Their effect is always to suppress (to average out, as we say) certain spin interactions and to let certain others survive. Whether or not a given type of spin interaction will be suppressed by a certain type of motion depends on the structure of the relevant Hamiltonian with respect to both its spin and space variables. Because of the crucial role the structures of nuclear spin Hamiltonians play in high-resolution NMR both in liquids and solids we devote the following chapter to their study.

Before beginning we want to mention, however, that there is at least one way of overcoming dipolar line broadening in solids that is *not* based on any kind of motional averaging. We stated above that it is the virtually complete randomness of the nuclear spin orientations under "normal circumstances" that causes the dipolar line broadening. Therefore, the conceptually simplest way to overcome this broadening is to line up all the spins neatly, in other words to polarize the sample highly. Samples with nuclear spin polarizations of the order of 90% can truly be called "abnormal." Nevertheless, the Saclay NMR group headed by Professors Abragam and Goldman has succeeded in pre-paring samples of CaF_2 with ^{19}F spin polarizations approaching or even exceeding 90%, and they have indeed observed a substantial narrowing of the ^{19}F NMR linewidth when the ^{19}F polarization approached 100%.

II. Nuclear Spin Hamiltonian

The nuclear spin Hamiltonian consists of a number of terms that describe physically different interactions of the nuclear spins. Some of these terms are related to the apparatus, and some to the physical properties of the sample. Some cause shifts, while others cause broadening of the NMR lines. As we have outlined in the introduction, it is the behavior of these Hamiltonian terms under motions (mostly, but not exclusively, rotations) in both spatial and spin coordinate spaces that enters basically all considerations about obtaining high resolution in spectra of both solid and liquid samples. In this chapter we study this behavior for the nuclear spin Hamiltonian terms relevant to diamagnetic, nonconducting substances. It is to this class of substances that, by and large, high-resolution NMR has been limited to date, although the so-called magic-angle sample-spinning high-resolution techniques have also been applied to metals.

A. Interactions of Nuclear Spins in Diamagnetic Nonconducting Substances

The nuclear spin Hamiltonian in diamagnetic nonconducting substances can be expressed in the following tabular form:

Term	Coupling of nuclear spins with
$h\mathcal{H} = h\mathcal{H}_z$	external static magnetic fields
$+ h\mathcal{H}_{rf}$	external rf magnetic fields
$+ h\mathcal{H}_{CS}$	induced magnetic fields originating from orbital motions of electrons
$+ h\mathcal{H}_Q$	electric field gradients
$+ h\mathcal{H}_{SR}$	magnetic moment associated with the molecular angular momentum
$+ h\mathcal{H}_D$	each other, directly through their magnetic dipole moments
$+ h\mathcal{H}_J$	each other, indirectly via electron spins

We shall now give a few comments on each of these terms.

5

1. ZEEMAN HAMILTONIAN $h\mathcal{H}_Z$

The static applied magnetic field $\mathbf{B}_{st}$ is usually chosen as $(0, 0, B_z)$, where we assume tacitly that it is constant in space. We shall have to drop this assumption occasionally. This choice of $\mathbf{B}_{st}$ leads to

$$h\mathcal{H}_Z = -\sum_i \mu_z^{\,i} B_z = -\hbar B_z \sum_i \gamma_n^{\,i} I_z^{\,i} = -\hbar \sum_i \omega_0^{\,i} I_z^{\,i}, \tag{2-1}$$

where $\mathbf{\mu}^i$ and $\gamma_n^{\,i}$ are, respectively, the magnetic moment and the magnetogyric ratio of the ith nucleus; i runs over all nuclei of the sample; $\omega_0^{\,i} = \gamma_n^{\,i} B_z$.

2. RADIO FREQUENCY HAMILTONIAN $h\mathcal{H}_{rf}$

The rf field is usually applied perpendicular to the static field. We choose it parallel to the x axis:

$$\mathbf{B}_{rf} = (B_1(t) \cos[\omega t + \varphi(t)], 0, 0). \tag{2-2}$$

This form of $\mathbf{B}_{rf}$ implies an rf irradiation that may be modulated in both its amplitude and phase, but that has a constant carrier frequency $\omega/2\pi$. Consequently, we have

$$h\mathcal{H}_{rf} = -\hbar B_1(t) \cos[\omega t + \varphi(t)] \sum_i \gamma_n^{\,i} I_x^{\,i}. \tag{2-3}$$

In multiple-resonance experiments $\mathbf{B}_{rf}$ consists of a sum of fields that differ, in particular, but not only, in their carrier frequencies ω^j.

3. CHEMICAL SHIFT OR NUCLEAR MAGNETIC SHIELDING HAMILTONIAN $h\mathcal{H}_{CS}$

This Hamiltonian can be written in the form

$$h\mathcal{H}_{CS} = \hbar \sum_i \gamma_n^{\,i} \mathbf{I}^i \cdot \mathbf{\sigma}^i \cdot \mathbf{B}, \tag{2-4}$$

where $\mathbf{\sigma}^i$ is a tensor of rank two, characteristic for each nuclear site; $-\mathbf{\sigma}^i \cdot \mathbf{B}$ is the magnetic field induced by the electrons at the site of the ith nucleus. Usually one restricts $\mathbf{B}$ to $\mathbf{B}_{st}$. We shall be concerned particularly with $\mathbf{\sigma}$ tensors in this volume.

4. THE QUADRUPOLAR HAMILTONIAN $h\mathcal{H}_Q$

This Hamiltonian can be written in the form

$$h\mathcal{H}_Q = \sum_i \frac{eQ^i}{6I^i(2I^i - 1)} \sum_{\alpha, \beta = 1}^{3} V_{\alpha\beta}^i [\tfrac{3}{2}(I_\alpha^i I_\beta^i + I_\beta^i I_\alpha^i) - \delta_{\alpha\beta}(\mathbf{I}^i)^2] \tag{2-5a}$$

$$= \sum_i \frac{eQ^i}{6I^i(2I^i - 1)} \mathbf{I}^i \cdot \mathbf{V}^i \cdot \mathbf{I}^i, \tag{2-5b}$$

where eQ^i and I^i are, respectively, the nuclear quadrupole moment and the nuclear spin quantum number of the ith nucleus; $V_{\alpha\beta}^i$ is the second (α, β) derivative of the electric potential at the site of the ith nucleus.

5. THE SPIN–ROTATION INTERACTION HAMILTONIAN

The relation

$$\hbar\mathscr{H}_{SR} = \hbar \sum_m \sum_i \mathbf{I}^i \cdot \mathbf{C}^{i,m} \cdot \mathbf{J}^m \qquad (2\text{-}6)$$

describes the coupling of the nuclear spins i of a molecule m with the magnetic moment associated with the angular momentum $\mathbf{J}^m$ of that molecule. Here i runs over all nuclei in the molecule m, and m runs over all molecules of the sample. Henceforth we shall drop the index m. The spin–rotation interaction tensor $\mathbf{C}^i$ depends, in principle, on the molecular rotational quantum number J: $\mathbf{C}^i = \mathbf{C}^i(J)$. This dependence, however, is weak and may often be neglected. $\mathbf{C}^i$ would be independent of J altogether if the nuclear geometrical configuration were independent of J.

$\mathscr{H}_Z$, $\mathscr{H}_{rf}$, $\mathscr{H}_{CS}$, $\mathscr{H}_Q$, and $\mathscr{H}_{SR}$ are sums of single-spin Hamiltonians. The next two terms are many-spin Hamiltonians. They couple, in principle, every spin of the sample with all others.

6. THE DIPOLAR HAMILTONIAN $\hbar\mathscr{H}_D$

This Hamiltonian is given by

$$\hbar\mathscr{H}_D = \sum_{i<k} \left(-\frac{\gamma_n^i \gamma_n^k \hbar^2}{r_{ik}^3} \right) \left[\frac{3(\mathbf{I}^i \cdot \mathbf{r}_{ik})(\mathbf{I}^k \cdot \mathbf{r}_{ik})}{r_{ik}^2} - \mathbf{I}^i \cdot \mathbf{I}^k \right] \qquad (2\text{-}7a)$$

$$= \sum_{i<k} (-2\gamma_n^i \gamma_n^k \hbar^2) \sum_{\alpha,\beta=1}^{3} [I_\alpha^i \cdot D_{\alpha\beta}^{ik} \cdot I_\beta^k], \qquad (2\text{-}7b)$$

where $\mathbf{r}_{ik}$ is the vector from nucleus i to nucleus k, and $|\mathbf{r}_{ik}| = r_{ik}$. The components $D_{\alpha\beta}^{ik}$ of the tensor $\mathbf{D}^{ik}$ are just the coefficients going with $I_\alpha^i I_\beta^k$.

7. THE INDIRECT SPIN–SPIN COUPLING TERM

This term may be written

$$\hbar\mathscr{H}_J = \hbar \sum_{i<k} \mathbf{I}^i \cdot \mathbf{J}^{ik} \cdot \mathbf{I}^k, \qquad (2\text{-}8)$$

where $\mathbf{J}^{ik}$ again is a tensor of rank two.

Let us call $\mathscr{H}_Z$ and $\mathscr{H}_{rf}$ external Hamiltonians because they depend, apart from nuclear properties (γ_n^i) only on external parameters $(B_z, B_1(t), \varphi(t), \ldots)$, which are under control of the experimenter. Consequently we call $\mathscr{H}_{CS}$, $\mathscr{H}_Q$,

$\mathscr{H}_{\text{SR}}$, $\mathscr{H}_{\text{D}}$, and $\mathscr{H}_{\text{J}}$ internal Hamiltonians. They all have a common structure. Disregarding summations over nuclei i or pairs of nuclei i, k they can be expressed as

$$\mathscr{H}_\lambda = C^\lambda \sum_{\alpha,\beta=1}^{3} I_\alpha R_{\alpha\beta}^\lambda A_\beta^{\ \lambda} = C^\lambda \sum_{\alpha,\beta=1}^{3} R_{\alpha\beta}^\lambda T_{\beta\alpha}^\lambda. \qquad (2\text{-}9)$$

The C^λ depend only on fundamental constants ($\hbar, e$) and properties of a definite nuclear state s ($I_s^i, \gamma_{sn}^i, Q_s^i$), which mostly but not always is the nuclear ground state. In this case we shall drop—as we did before—the index s. In particular, the C^λ are constants with respect to rotations:

$$C^\lambda = \begin{cases} 1, & \\[4pt] \gamma_n^{\ i}, & \\[4pt] \dfrac{eQ^i}{6I^i(2I^i-1)\hbar}, & \\[4pt] -2\gamma_n^{\ i}\gamma_n^{\ k}\hbar, & \end{cases} \text{for} \quad \lambda = \begin{cases} \text{SR, J,} & (2\text{-}10\text{a}) \\[4pt] \text{CS,} & (2\text{-}10\text{b}) \\[4pt] \text{Q,} & (2\text{-}10\text{c}) \\[4pt] \text{D.} & (2\text{-}10\text{d}) \end{cases}$$

The differences between C^{CS} and C^{SR}, and between C^{D} and C^{J} have their origin in conventions of defining the $\mathscr{H}^\lambda$, and not in physics!

The $T_{\beta\alpha}^\lambda$ are dyadic products constructed from two vectors, one of which is always a nuclear spin vector, whereas the other one ($\mathbf{A}^\lambda$) can be the same nuclear spin vector ($\lambda = \text{Q}$), another nuclear spin vector ($\lambda = \text{D, J}$), the external magnetic field ($\lambda = \text{CS}$), or the molecular angular momentum vector $\mathbf{J}$ ($\lambda = \text{SR}$).

The $\mathbf{R}^\lambda (\mathbf{C}^i, \mathbf{V}^i, \boldsymbol{\sigma}^i, \mathbf{D}^{ik}, \mathbf{J}^{ik})$ depend on a definite electronic state—mostly but not exclusively—the electronic ground state of the molecule or crystal, and parametrically on nuclear geometrical configurations, and vibrational and rotational states of the molecules.

All $\mathbf{R}^\lambda$ are tensors of rank two. Hence they can be decomposed into their irreducible constituents with respect to the full 3d rotation group, O3: $\mathbf{R} = \mathbf{R}^{(0)} + \mathbf{R}^{(1)} + \mathbf{R}^{(2)}$.

$\mathbf{R}^{(0)} = \frac{1}{3}\text{Tr } \mathbf{R}\mathbb{1} = R\mathbb{1}$ is the isotropic constituent. $\mathbb{1}$ is the unit dyadic. The components of the traceless antisymmetric ($\mathbf{R}^{(1)}$) and traceless symmetric ($\mathbf{R}^{(2)}$) constituents are given, respectively, by

$$R_{\alpha\beta}^{(1)} = \tfrac{1}{2}(R_{\alpha\beta} - R_{\beta\alpha}) \quad \text{and} \quad R_{\alpha\beta}^{(2)} = \tfrac{1}{2}(R_{\alpha\beta} + R_{\beta\alpha}) - R\,\delta_{\alpha\beta}.$$

$\mathbf{C}$, $\boldsymbol{\sigma}$, and $\mathbf{J}$ can contain, in principle, all three constituents. However, only the isotropic and traceless symmetric parts are measurable by spectroscopic techniques (see Chapter III, Section D). $\mathbf{D}$ and $\mathbf{V}$ are traceless symmetric by their very nature. In what follows we shall therefore ignore eventual antisymmetric constituents and shall treat all $\mathbf{R}^\lambda$ as symmetric.

There exist principal axes systems (PAS)—different from each other in general—for all $\mathbf{R}^\lambda$. These PAS's are fixed in some molecular or crystal frame and eventually they move together with the molecule or crystal. The $\mathbf{R}^\lambda$-tensors assume very simple forms in their respective PAS's: They become diagonal. The diagonal elements in the PAS are called principal components, R_{XX}, R_{YY}, R_{ZZ}.

The following convention is chosen to label the axes:

$$|R_{ZZ} - R| \geqslant |R_{XX} - R| \geqslant |R_{YY} - R|. \qquad (2\text{-}11)$$

The quantities between the absolute signs, $R_{\alpha\alpha} - R$, are the principal components of $\mathbf{R}^{(2)}$.

It is often convenient to introduce instead of the three parameters $R_{\alpha\alpha}$ three new parameters, one of which is

$$R = \tfrac{1}{3}\operatorname{Tr}\mathbf{R}, \qquad (2\text{-}12)$$

and the other two, δ and η, are defined by

$$\delta = R_{ZZ} - \tfrac{1}{3}\operatorname{Tr}\mathbf{R} = R_{ZZ} - R, \qquad (2\text{-}13)$$

$$\eta = \frac{R_{YY} - R_{XX}}{R_{ZZ} - \tfrac{1}{3}\operatorname{Tr}\mathbf{R}} = \frac{R_{YY} - R_{XX}}{\delta}. \qquad (2\text{-}14)$$

In terms of the $R_{\alpha\alpha}$ and of the new parameters R, δ, η we have in the PAS

$$\mathbf{R}(\text{PAS}) = \begin{bmatrix} R_{XX} & & \\ & R_{YY} & \\ & & R_{ZZ} \end{bmatrix} = R\mathbb{1} + \delta \begin{bmatrix} -\tfrac{1}{2}(1+\eta) & & \\ & -\tfrac{1}{2}(1-\eta) & \\ & & 1 \end{bmatrix}. \qquad (2\text{-}15)$$

$R^D = R^Q = 0$. In high-resolution NMR in fluids $R^{CS} = \tfrac{1}{3}\operatorname{Tr}\sigma = \sigma$, and $R^{J,ik} = \tfrac{1}{3}\operatorname{Tr}\mathbf{J}^{ik} = J^{ik}$ are called, respectively, "chemical shifts" and (scalar) "coupling constants." $C = \tfrac{1}{3}\operatorname{Tr}\mathbf{C} \neq 0$, in general; however, it causes at most line broadening but never line shifts or line splittings in NMR spectra of condensed matter.

$\delta^{D,ik} = r_{ik}^{-3}$. The quadrupole resonance community calls $\delta^{Q,i}$ "field gradient" (at the site of nucleus i) and denotes it by eq^i. $\tfrac{3}{2}\delta^{CS,i} = \sigma_{ZZ}^i - \tfrac{1}{2}(\sigma_{XX}^i + \sigma_{YY}^i)$ may be called chemical shift or nuclear magnetic shielding anisotropy of the ith nucleus. For axially symmetric shielding tensors with $\sigma_{XX} = \sigma_{YY} = \sigma_\perp$ and $\sigma_{ZZ} = \sigma_\parallel$ the total anisotropy range is $\Delta\sigma = \sigma_\parallel - \sigma_\perp = \tfrac{3}{2}\delta^{CS}$. $\delta^{J,ik}$ and $\delta^{SR,i}$ have not yet been given special names. η^λ ($\lambda = D, Q, CS, J, SR$) is called the asymmetry parameter of the corresponding tensor. $\eta^\lambda = 0$ means, obviously, that $\mathbf{R}^\lambda$ is axially symmetric. The dipolar interaction between two magnetic moments is axially symmetric about the line interconnecting the moments, hence $\eta^D = 0$.

B. Behavior of Internal Hamiltonians under Rotations

All techniques used for *selective averaging* in NMR rely on rotations of one kind or another of our internal Hamiltonians. The behavior under rotations of any physical property can be studied most easily when this property is expressed in terms of components of irreducible spherical tensor operators, denoted by T_{lm} or R_{lm}. The behavior of irreducible spherical tensor operators under rotations is well known. The books by Edmonds,[1] Rose,[2] and Brink and Satchler[3] are recommended for an introduction to spherical tensor operators.

The equivalent of Eq. (2-9) in irreducible spherical tensor operator calculus is

$$\mathcal{H}_\lambda = C^\lambda \sum_l \sum_{m=-l}^{l} (-1)^m R_{l,-m}^\lambda T_{lm}^\lambda, \tag{2-16}$$

where the R_{lm}^λ derive from the $R_{\alpha\beta}^\lambda$ and the T_{lm}^λ from the $T_{\alpha\beta}^\lambda$. If we start with symmetric 2nd-rank cartesian tensors $R_{\alpha\beta}^\lambda$, only R_{lm}^λ's with $l = 0, 2$ will be nonzero. If we consider the $\mathbf{R}^\lambda$ tensors in their principal axes systems, only components with $m = 0, \pm 2$ are nonzero.

We shall denote the components of irreducible spherical tensor operators R_{lm} in their respective PAS's by ρ_{lm}. The relations between the nonzero ρ_{lm}'s and the $R_{\alpha\alpha}$ (or R, δ, and η) are

$$\rho_{00} = \tfrac{1}{3} \operatorname{Tr} \mathbf{R} = R, \tag{2-17}$$

$$\rho_{20} = \sqrt{\tfrac{3}{2}}(R_{ZZ} - \tfrac{1}{3} \operatorname{Tr} \mathbf{R}) = \sqrt{\tfrac{3}{2}}\delta, \tag{2-18}$$

$$\rho_{2\pm2} = \tfrac{1}{2}(R_{YY} - R_{XX}) = \tfrac{1}{2}\eta\delta. \tag{2-19}$$

We shall express the T_{lm}^λ (derived from the $T_{\beta\alpha}^\lambda = I_\alpha A_\beta^\lambda$) in the laboratory frame of reference because the $I_\alpha, A_\beta^\lambda$ are basically linked to that frame[4]; therefore, we must express the R_{lm}^λ also in the laboratory frame.

[1] A. R. Edmonds, "Angular Momentum in Quantum Mechanics." Princeton Univ. Press, Princeton, New Jersey, 1957.

[2] M. E. Rose, "Elementary Theory of Angular Momentum." Wiley, New York, 1957.

[3] D. M. Brink and G. R. Satchler, "Angular Momentum." Oxford Univ. Press (Clarendon), London and New York, 1971.

[4] This is obvious for the components of the magnetic fields, both static and oscillatory, but it is also true for the spin components. The spin components are the dynamic variables in NMR experiments. It is their time evolution that we observe using laboratory equipment and hence naturally in a laboratory frame of reference. The role of the $\mathbf{D}, \mathbf{V}, \mathbf{J}, \sigma$, and $\mathbf{C}$ tensors is quite different: they are parameters that govern the time evolution of the dynamic variables, i.e., of the spin system. The time evolution of the spin system carries information about the parameters in which we are ultimately interested.

Being irreducible spherical tensor operators, they may be expressed in the laboratory frame in terms of the ρ^λ_{lm}, and of the Wigner rotation matrices $\mathscr{D}^l_{m'm}(\alpha^\lambda, \beta^\lambda, \gamma^\lambda)$ (see, e.g., Edmonds[1]):

$$R^\lambda_{lm} = \sum_{m'} \mathscr{D}^l_{m'm}(\alpha^\lambda, \beta^\lambda, \gamma^\lambda)\, \rho^\lambda_{lm'}, \qquad (2\text{-}20)$$

where $\alpha^\lambda, \beta^\lambda, \gamma^\lambda$ (the triple of which we shall denote by Ω^λ) are the Euler angles by which the laboratory frame can be brought into coincidence with the λth principal axes system. $\mathscr{D}^l_{m'm} = \delta_{m'm}$ for $\Omega = 0$, i.e., when the laboratory frame and the principal axes system coincide. In such a situation $R^\lambda_{lm} = \rho^\lambda_{lm}$, as it should.

$\mathscr{D}^0_{00}$ is a constant (unity). The only Wigner matrix actually needed in our context is the $\mathscr{D}^2_{m'm}$ matrix given in Table 2-1.

We can now turn to the last factors in Eq. (2-9) or (2-16). The elements of the dyadic product $T^\lambda_{\beta\alpha} = I_\alpha A_\beta{}^\lambda$ form a basis for a nine-dimensional, reducible representation of the rotation group. Its reduction leads to the desired irreducible bases sets T^λ_{lm}. The reduction can be carried out once and for all for dyadic products. The results are listed in Table 2-2. We omit the 3d bases, because their counterparts in Eq. (2-16), the R_{1m}'s vanish (or are ignored). The following definitions have been used in Table 2-2:

$$I_\pm = I_x \pm iI_y, \qquad (2\text{-}21)$$

$$I_{+1} = -\frac{1}{\sqrt{2}} I_+, \qquad I_0 = I_z, \qquad I_{-1} = +\frac{1}{\sqrt{2}} I_-. \qquad (2\text{-}22)$$

Similar definitions have been used for the components of $\mathbf{B}_{st}$. The idea of these definitions is to express also vector components in terms of irreducible spherical tensor components, thus, e.g., $I_{\pm 1} = T_{1\pm 1}$, $I_0 = T_{10}$.

Equations (2-16)–(2-20) and Tables 2-1 and 2-2 enable us now to express all internal Hamiltonians in terms of rotational invariants (C^λ), irreducible spherical tensor operators ($\rho^\lambda_{lm}, T^\lambda_{lm}$), and elements of the Wigner matrix $\mathscr{D}^2_{m'm}$. We shall discuss explicitly two examples ($\mathscr{H}_D$ and $\mathscr{H}_{CS}$), which are of the greatest importance in the field of high-resolution NMR in solids.

1. DIPOLAR HAMILTONIAN $\mathscr{H}_D$

We recall

$$C^{D, ik} = -2\gamma_n{}^i\gamma_n{}^k\hbar, \qquad \rho^{D, ik}_{20} = \sqrt{\tfrac{3}{2}}\, \delta^{D, ik} = \sqrt{\tfrac{3}{2}}\, r_{ik}^{-3}, \qquad \rho^D_{2m'} = 0, \quad \text{for} \quad m' \neq 0.$$

Equation (2-20) yields

$$R^{D, ik}_{2m} = \mathscr{D}^2_{0m}(\Omega^{D, ik})\sqrt{\tfrac{3}{2}}\, r_{ik}^{-3}.$$

TABLE 2-1

WIGNER ROTATION MATRIX $\mathscr{D}^2_{m'm}(\alpha,\beta,\gamma)$

m'			m		
	2	1	0	-1	-2
2	$\dfrac{(1+\cos\beta)^2}{4}e^{2i(\alpha+\gamma)}$	$-\dfrac{1+\cos\beta}{2}\sin\beta\,e^{i(2\alpha+\gamma)}$	$\sqrt{\tfrac{3}{8}}\sin^2\beta\,e^{2i\alpha}$	$-\dfrac{1-\cos\beta}{2}\sin\beta\,e^{i(2\alpha-\gamma)}$	$\dfrac{(1-\cos\beta)^2}{4}e^{2i(\alpha-\gamma)}$
1	$\dfrac{1+\cos\beta}{2}\sin\beta\,e^{i(2\gamma+\alpha)}$	$\left[\cos^2\beta-\dfrac{1-\cos\beta}{2}\right]e^{i(\alpha+\gamma)}$	$-\sqrt{\tfrac{3}{8}}\sin2\beta\,e^{i\alpha}$	$\left[\dfrac{1+\cos\beta}{2}-\cos^2\beta\right]e^{i(\alpha-\gamma)}$	$-\dfrac{1-\cos\beta}{2}\sin\beta\,e^{i(\alpha-2\gamma)}$
0	$\sqrt{\tfrac{3}{8}}\sin^2\beta\,e^{2i\gamma}$	$\sqrt{\tfrac{3}{8}}\sin2\beta\,e^{i\gamma}$	$\dfrac{3\cos^2\beta-1}{2}$	$-\sqrt{\tfrac{3}{8}}\sin2\beta\,e^{-i\gamma}$	$\sqrt{\tfrac{3}{8}}\sin^2\beta\,e^{-2i\gamma}$
-1	$\dfrac{1-\cos\beta}{2}\sin\beta\,e^{i(2\gamma-\alpha)}$	$\left[\dfrac{1+\cos\beta}{2}-\cos^2\beta\right]e^{i(\gamma-\alpha)}$	$\sqrt{\tfrac{3}{8}}\sin2\beta\,e^{-i\alpha}$	$\left[\cos^2\beta-\dfrac{1-\cos\beta}{2}\right]e^{-i(\alpha+\gamma)}$	$\dfrac{1+\cos\beta}{2}\sin\beta\,e^{-i(\alpha+2\gamma)}$
-2	$\dfrac{(1-\cos\beta)^2}{4}e^{2i(\gamma-\alpha)}$	$\dfrac{1-\cos\beta}{2}\sin\beta\,e^{i(\gamma-2\alpha)}$	$\sqrt{\tfrac{3}{8}}\sin^2\beta\,e^{-2i\alpha}$	$\dfrac{1+\cos\beta}{2}\sin\beta\,e^{-i(2\alpha+\gamma)}$	$\dfrac{(1+\cos\beta)^2}{4}e^{-2i(\alpha+\gamma)}$

TABLE 2-2

ONE- AND FIVE-DIMENSIONAL IRREDUCIBLE STANDARDIZED BASES SETS T_{lm} OF O3 CONTAINED IN $I_\alpha A_\beta^\lambda$

Interaction	λ	$I_\alpha A_\beta^\lambda$	T_{00}	T_{20}	$T_{2\pm1}$	$T_{2\pm2}$
Shielding.	CS	$I_\alpha^i B_\beta$	$I_0^i B_0$	$\sqrt{\dfrac{2}{3}}\,I_0^i B_0$	$\dfrac{1}{\sqrt{2}} I_{\pm1}^i B_0$	0
Spin rotation	SR	$I_\alpha^i J_\beta$	$\mathbf{I}^i\cdot\mathbf{J}$	$\dfrac{1}{\sqrt{6}}[3I_0^iJ_0 - \mathbf{I}^i\cdot\mathbf{J}]$	$\dfrac{1}{\sqrt{2}}[I_{\pm1}^iJ_0 + I_0^iJ_{\pm1}]$	$I_{\pm1}^iJ_{\pm1}$
Quadrupole	Q	$I_\alpha^iI_\beta^i$	$(I^i)^2$	$\dfrac{1}{\sqrt{6}}[3(I_0^i)^2 - (I^i)^2]$	$\dfrac{1}{\sqrt{2}}[I_{\pm1}^iI_0^i + I_0^iI_{\pm1}^i]$	$(I_{\pm1}^i)^2$
Dipole $\left.\begin{array}{c}\text{D}\\\text{J}\end{array}\right\}$ Indirect spin–spin		$I_\alpha^iI_\beta^k$	$\mathbf{I}^i\cdot\mathbf{I}^k$	$\dfrac{1}{\sqrt{6}}[3I_0^iI_0^k - \mathbf{I}^i\cdot\mathbf{I}^k]$	$\dfrac{1}{\sqrt{2}}[I_{\pm1}^iI_0^k + I_0^iI_{\pm1}^k]$	$I_{\pm1}^iI_{\pm1}^k$

Inserting $R_{2m}^{D,ik}$ and $C^{D,ik}$ into Eq. (2-16) leads to

$$\mathscr{H}^{D,ik} = -\sqrt{6}\,\gamma_n{}^i\gamma_n{}^k \hbar r_{ik}^{-3} \sum_m (-1)^m \mathscr{D}_{0,-m}^2(\Omega^{ik}) T_{2m}^{D,ik}. \qquad (2\text{-}23)$$

Now the $\mathscr{D}_{0m}^2$ are—apart from a constant factor—identical with the spherical harmonics Y_{2m}, which behave under rotations exactly as the R_{lm}:

$$\mathscr{D}_{0m}^2(\alpha,\beta,\gamma) = (4\pi/5)^{1/2} Y_{2m}(\beta\alpha). \qquad (2\text{-}24\text{a})$$

Similarly,

$$\mathscr{D}_{m'0}^2(\alpha,\beta,\gamma) = (-1)^{m'}(4\pi/5)^{1/2} Y_{2m'}(\beta,\gamma). \qquad (2\text{-}24\text{b})$$

The Y_{lm} are normalized such that

$$\int_{\text{sphere}} Y_{lm} Y_{l'm'}^* \, d\mathbf{s} = \delta_{ll'}\,\delta_{mm'}, \qquad (2\text{-}25)$$

where $d\mathbf{s}$ is the surface element on the sphere.

Note that the $\mathscr{D}_{0m}^2$'s actually do not depend on γ; note also that $\beta^{D,ik}$ and $\alpha^{D,ik}$ are identical, respectively, with the polar angles ϑ^{ik} and φ^{ik} of the internuclear vector $\mathbf{r}^{ik}$ in the laboratory frame.

Inserting Eq. (2-24) into Eq. (2-23) and reintroducing the indices k and i, we have finally[5]

$$\mathscr{H}_D = -2(6\pi/5)^{1/2}\hbar \sum_{i<k} \gamma_n{}^i\gamma_n{}^k r_{ik}^{-3} \sum_m (-1)^m Y_{2,-m}(\vartheta^{ik},\varphi^{ik}) T_{2m}^{D,ik}. \qquad (2\text{-}26)$$

The $T_{2m}^{D,ik}$ are given explicitly in the last row of Table 2-2. This form of $\mathscr{H}_D$ is the most useful one not only for discussing high-resolution NMR in solids but also for, e.g., discussing nuclear magnetic relaxation phenomena.[6]

2. Shielding Hamiltonian $\mathscr{H}_{CS}$

In irreducible spherical tensor notation the shielding Hamiltonian of a given nucleus assumes the form

$$\mathscr{H}_{CS} = \gamma_n \sum_{l=0,2} \sum_{m=-l}^{+l} (-1)^m R_{l,-m}^{CS} T_{lm}^{CS}. \qquad (2\text{-}27)$$

On the right-hand side of the following equations we shall drop the index CS.

---

[5] Equation (2-26) differs from Eq. (2) of Haeberlen and Waugh[6] by a factor of 2. This is due to a different definition of the T_{lm}^D. In Haeberlen and Waugh[6] the definition of Slichter[7] has been used.

[6] U. Haeberlen and J. S. Waugh, *Phys. Rev.* **185**, 420 (1969).

[7] C. P. Slichter, "Principles of Magnetic Resonance," p. 168. Harper, New York, 1964.

Introducing the principal axes shielding components ρ_{lm} into Eq. (2-27) gives

$$\mathcal{H}_{CS} = \sum_{l=0,2} \sum_{m=-l}^{+l} (-1)^m T_{lm} \gamma_n \sum_{m'} \mathcal{D}^l_{m',-m} \rho_{lm'} \tag{2-28a}$$

$$= \gamma_n T_{00} \rho_{00} \tag{2-28b}$$

$$+ \gamma_n \sum_{m=-2}^{+2} (-1)^m T_{2m} \sum_{m'} \mathcal{D}^2_{m',-m} \rho_{2m'} \qquad\qquad \underline{\quad l \quad\quad m \quad}$$

$$= \omega_0 \sigma I_0 \qquad\qquad\qquad\qquad\qquad\qquad\qquad\qquad\qquad 0 \quad\quad 0$$

$$+ \omega_0 \left\{ \sqrt{\tfrac{2}{3}} I_0 \delta \left[\sqrt{\tfrac{3}{2}} \mathcal{D}^2_{00} + \tfrac{\eta}{2}(\mathcal{D}^2_{20} + \mathcal{D}^2_{-20}) \right] \right. \qquad 2 \quad\quad 0$$

$$- \tfrac{1}{\sqrt{2}} I_{-1} \delta \left[\sqrt{\tfrac{3}{2}} \mathcal{D}^2_{01} + \tfrac{\eta}{2}(\mathcal{D}^2_{21} + \mathcal{D}^2_{-21}) \right] \qquad 2 \quad\; -1$$

$$\left. - \tfrac{1}{\sqrt{2}} I_{+1} \delta \left[\sqrt{\tfrac{3}{2}} \mathcal{D}^2_{0,-1} + \tfrac{\eta}{2}(\mathcal{D}^2_{2,-1} + \mathcal{D}^2_{-2,-1}) \right] \right\}. \qquad 2 \quad\; +1$$

$$\tag{2-28c}$$

There are no terms $l = 2$, $m = \pm 2$, because the $T_{2\pm 2}$ involved are zero. The $m = 0$ terms, i.e., the secular terms, are of the greatest importance. We give three more formulations for them [Eqs. (2-29)–(2-31)].

Inserting Eq. (2-24b) into Eq. (2-28c) gives

$$\mathcal{H}_{CS,\,secular} = \omega_0 I_0 \{\sigma + (8\pi/15)^{1/2}\delta [(3/2)^{1/2} Y_{20} + \tfrac{1}{2}\eta (Y_{22} + Y_{2,-2})]\}. \tag{2-29}$$

The arguments of the spherical harmonics are $\beta, \gamma = \beta^{CS}, \gamma^{CS}$.

By inserting the appropriate trigonometric expressions for the Y_{2m}'s we get

$$\mathcal{H}_{CS,\,secular} = \omega_0 I_0 \left\{ \sigma + \delta \left[\frac{3\cos^2\beta - 1}{2} + \frac{1}{2}\eta \sin^2\beta \cos 2\gamma \right] \right\}. \tag{2-30}$$

By direct evaluation one may convince oneself that

$$\mathcal{H}_{CS,\,secular} = \omega_0 I_0 \sigma_{zz} \tag{2-31}$$

where σ_{zz} is the zz-shielding component in the laboratory frame.

While admittedly Eq. (2-31) looks much simpler than Eqs. (2-29) and (2-30), the latter have the definite advantage that they display much clearer (i) the rotational behavior of $\mathcal{H}_{CS,\,secular}$ and (ii) the parameters on which it depends. Note that $\mathcal{H}_{CS,\,secular}$ does not depend on the Euler angle α. This reflects the fact that the NMR experiment is invariant under rotations about the applied field. Still another expression for $\mathcal{H}_{CS,\,secular}$ will be needed and given in Chapter III, (Eq. 3-6).

III. Manifestations of Nuclear Magnetic Shielding in NMR Spectra of Solids

A. General Remarks

We have already stated that measuirng the NMR parameters σ^i, $\mathbf{D}^{ik}$, $\mathbf{V}^i$, etc., is one of the principal goals of experimental NMR. There are three main factors governing the experimenter's access to these parameters:

(1)　the relative size of the internal Hamiltonians,

(2)　the averaging processes operative—intentionally or unintentionally—in both ordinary $\mathbf{r}$ and spin spaces,

(3)　the manner in which the internal Hamiltonians manifest themselves in NMR spectra, particularly in NMR spectra of solids.

In many cases $\mathscr{H}_D$ greatly exceeds other internal Hamiltonians, in particular $\mathscr{H}_{CS}$. It then renders σ inaccessible to measurement. In isotropic liquids, fast random molecular motions average out $\mathscr{H}_D$, but likewise they average out all anisotropic constituents of $\mathscr{H}_{CS}$. Only the isotropic constituent of σ becomes measurable. The major achievement of selective averaging by multiple-pulse sequences is that it provided a technique by which it is possible to suppress $\mathscr{H}_D$ in solids by an averaging process in spin space that leaves alive all constituents of $\mathscr{H}_{CS}$. By this technique the full σ tensors become accessible to measurement.

In this chapter we discuss the types of spectra encountered when $\mathscr{H}_{CS}$ either is the dominant part of the total spin Hamiltonian or when it is made the dominant part by selectively suppressing the homo- and heteronuclear dipolar spin–spin coupling terms.

Let us consider a perfect single crystal and let us recall two definitions:

Two nuclei are called crystallographically equivalent if they are related by any one of the symmetry elements of the space group of the crystal.

17

Two nuclei are called magnetically equivalent if they are related by one of the translation and/or inversion elements of the space group.[8] These latter symmetry elements constitute an invariant subgroup of the space group.

Using these definitions, we may state:

(1)   The tensorial parameters σ and $\mathbf{V}$ are identical for magnetically equivalent nuclei, i.e., $\sigma^i \equiv \sigma^k$, $\mathbf{V}^i \equiv \mathbf{V}^k$ for magnetically equivalent nuclei i and k.

(2) σ and $\mathbf{V}$ of crystallographically equivalent nuclei i and k are related by the rotation–reflection parts of the symmetry operations by which the nuclei i and k are related. The invariance of σ and $\mathbf{V}$ against inversions is a consequence of the symmetry of these tensors ($\sigma_{\alpha\beta} = \sigma_{\beta\alpha}$; $V_{\alpha\beta} = V_{\beta\alpha}$, or alternatively, $\rho_{2\pm1}^{\text{CS,Q}} = 0$). Effects of hypothetical nonvanishing constituents of σ are discussed below.

(3)   The NMR spectra of idealized samples, in which only single-particle Hamiltonians such as $\mathscr{H}_{\text{CS}}$ and $\mathscr{H}_{\text{Q}}$ are operative, are independent superpositions of the spectra from magnetically nonequivalent nuclei.

Each partial spectrum consists of N_I sharp lines, where N_I depends on the spin quantum number I and the ratio $\hbar\omega_0{}^i/e^2 q^i Q^i$. For $\hbar\omega_0{}^i/e^2 q^i Q^i \gg 1$, $N_I = 2I$. For $I = \frac{1}{2}$, N_I is just 1. Line splitting, and eventually line broadening, is caused by the many-particle interactions $\mathscr{H}_{\text{D}}$ and $\mathscr{H}_{\text{J}}$. We disregard them in this chapter.

If there are N_m magnetically nonequivalent sites for a given isotope in a single crystal, its (idealized) NMR spectrum consists of $N_I \times N_m$ sharp lines. In special orientations of the crystal with respect to $\mathbf{B}_{\text{st}}$ some of the lines will coalesce. In any case the number of lines is limited, although often it is not very small.

Let us consider a crystal with N_C crystallographically equivalent but magnetically nonequivalent sites for an isotope with, say, spin $I = \frac{1}{2}$. These sites contribute N_C lines to the spectrum. Of course, we would like to know which of these lines arises from which site. Now it is very important to realize that it is impossible in principle to decide—on the basis of complete structural and NMR information—which of these lines must be assigned to which site. In other words, there is no symmetry argument on which an assignment could be based. With the procedures to be described in the following section we

---

[8] Note that neither crystallographically nor magnetically equivalent nuclei are automatically equivalent nuclei according to the definition used in high-resolution NMR in fluids, where equivalence is defined with respect to molecular symmetry, and scalar couplings J^{ik} are taken into account.[9]

[9] A. Abragam, "The Principles of Nuclear Magnetism," Chapter XI, p. 480. Oxford Univ. Press (Clarendon), London and New York, 1961.

can, in principle, determine a $\boldsymbol{\sigma}$ tensor for each of the N_C lines. These tensors transform among themselves under the symmetry operations by which the crystallographically equivalent sites are related.

Crystallographically nonequivalent sites can, in general, be distinguished: The corresponding $\boldsymbol{\sigma}$ tensors cannot be brought into exact coincidence by any one of the symmetry operations of the space group. There is, however, still the question of which set of lines (or $\boldsymbol{\sigma}$ tensors) belongs to which set of crystallographically equivalent sites. Again there is no symmetry argument that allows an assignment of the different sets of related lines to the corresponding sets of crystallographically equivalent sites if these sites are general sites. Special sites, however, can be distinguished from general sites. Sites on n-fold axes, as well as on screw axes, mirror and glide planes, and inversion centers are called special sites.

The lines from two sites related by, e.g., a mirror plane coincide when $\mathbf{B}_{st}$ falls into that mirror plane. This line then has double intensity in comparison to a line from a site on the mirror plane. More important, however, the former line splits when $\mathbf{B}_{st}$ moves out of the mirror plane, whereas the latter does not.

For magnetic resonance experiments planes perpendicular to n-fold axes (screw axes) are equivalent to mirror planes. Lines from sites on inversion centers can be distinguished from other lines again by their smaller intensity.

In summary, symmetry arguments are very valuable for handling the assignment problem, but often they do not lead to its complete solution. Assignments must then be based on additional information, for instance, NMR data from related compounds, or on theoretical grounds. Such assignments are usually highly reliable but inevitably they incorporate some kind of assumption.

B. Single Crystals Available

We start with a case in which all sites are magnetically equivalent for the isotope in which we are interested. For simplicity we suppose $I = \frac{1}{2}$. There is only one line in the spectrum. We ask ourselves how we can determine the desired shielding parameters from the spectral position of this line. Obviously we must render the line position dependent on an external parameter. The orientation of the crystal with respect to the external magnetic field plays the role of this external parameter.

In the approximation—usually an excellent one—in which we restrict ourselves to secular terms of the internal Hamiltonian [terms with $m = 0$ in Eq. (2-16)] the position of the one line in the spectrum is governed by

$$\mathscr{H}^i_{CS,\,secular} = \gamma_n I_0 B_0 \{\sigma^i + \sqrt{\tfrac{2}{3}}\, R^{CS,\,i}_{20}\} = \omega_0 I_0 \{\sigma^i + \sqrt{\tfrac{2}{3}}\, \sigma^i_{20}\}$$

[cf. Eq. (2-27)]. To save indices we set $R_{2m}^{CS, i} = \sigma_{2m}^i$. The upper Zeeman level $(m_I = -\frac{1}{2})$ is shifted by an amount $-\frac{1}{2}\hbar\omega_0\{\sigma^i + \sqrt{\frac{2}{3}}\sigma_{20}^i\}$, and the lower $(m_I = +\frac{1}{2})$ by the same amount in the opposite direction. Hence the position ω^i of the line relative to the position ω^b of the line of the bare nucleus $(\sigma^b \equiv 0)$ is given in units of 2π Hz by

$$\omega^i - \omega^b = -\omega_0\{\sigma^i + \sqrt{\tfrac{2}{3}}\sigma_{20}^i\} = -\omega_0\sigma_{zz}^i. \tag{3-1}$$

The position of the line of the bare nucleus is not known. Therefore, in practice, we do not reference ω^i against ω^b, but against ω^{ref}, which is the line position of a reference sample for which σ_{20}^i vanishes for one reason or another (e.g., motional averaging or cubic crystal symmetry).

$$\omega^i - \omega^{ref} = -\omega_0\{\sigma^i - \sigma^{ref} + \sqrt{\tfrac{2}{3}}\sigma_{20}^i\} \tag{3-2}$$

is all that we can measure in an NMR experiment. The scalar shielding parameter that we can measure—in principle, not only in practice—is not σ^i, but $\sigma^i - \sigma^{ref}$.

In Eq. (2-30) we have expressed σ_{20}^i in terms of δ^i, η^i and β^i, γ^i. Recall that β^i and γ^i are the Euler angles by which the laboratory system of coordinates (henceforth denoted by LABS) can be brought into coincidence with the PAS of σ^i. Although β^i and γ^i are parameters on which $\omega^i - \omega^{ref}$ directly depends, they are not the parameters in which we are ultimately interested.

The parameters in which we are ultimately interested are $\sigma^i - \delta^{ref}, \delta^i, \eta^i$, and the Euler angles $\alpha_i', \beta_i', \gamma_i'$ ($\equiv \Omega_i'$), which relate some arbitrarily chosen, crystal fixed orthogonal axes system (CRS) with the shielding principal axes system.

What is needed, therefore, is an expression for σ_{20}^i in terms of $\delta^i, \eta^i, \Omega_i'$, and the Euler angles $\alpha'', \beta'', \gamma''$ ($\equiv \Omega''$), which relate the laboratory and the crystal fixed frames. Such an expression is obtained easily by applying Eq. (2-20) twice:

$$\sigma_{20}^i(\text{LABS}) = \sum_{m'}\mathscr{D}_{m'0}^2(\Omega'')\sigma_{2m'}^i(\text{CRS}) = \sum_{m'}\mathscr{D}_{m'0}^2(\Omega'')\sum_{m''}\mathscr{D}_{m''m'}^2(\Omega_i')\rho_{2m''}^{CS, i}$$
$$= \sum_{m'}\mathscr{D}_{m'0}^2(\Omega'')\delta^i\left[\sqrt{\tfrac{3}{2}}\mathscr{D}_{0m'}^2(\Omega_i') + \tfrac{1}{2}\eta^i[\mathscr{D}_{2m'}^2(\Omega_i') + \mathscr{D}_{-2m'}^2(\Omega_i')]\right]. \tag{3-3}$$

Recall that the $\mathscr{D}_{m'0}^2$ actually do not depend on α''. In fact, ω^i must not depend on α'' since the line position must be invariant under rotations of the crystal about the external magnetic field.

Often it is very useful to express the relation between the LABS and the CRS not by the Euler angles $\alpha'', \beta'', \gamma''$—which are always hard to visualize, and one of which does not enter the result in any case—but by the polar angles ϑ, φ, which specify the direction of the external magnetic field in the crystal fixed frame. β'', γ'' and ϑ, φ are related by

$$\beta'' = \vartheta, \tag{3-4a}$$

$$\gamma'' = \pi - \varphi \tag{3-4b}$$

(see, e.g., Fig. 1.1 of Edmonds[1]).

Using Eqs. (2-24b), (3-4a), and (3-4b), and well-known symmetry relations among the spherical harmonics, we get

$$\mathscr{D}^2_{m'0}(\Omega'') = (-1)^{m'}(4\pi/5)^{1/2}Y_{2m'}(\beta'', \gamma'')$$

$$= (-1)^{m'}(4\pi/5)^{1/2}Y_{2m'}(\vartheta, \pi - \varphi)$$

$$= (4\pi/5)^{1/2}Y_{2,-m'}(\vartheta, \varphi). \tag{3-5}$$

Inserting Eqs. (3-5) and (3-3) into Eq. (3-2) gives the line position in terms of the desired parameters:

$$\omega^i - \omega^{\text{ref}} = -\omega_0 \left\{ \sigma^i - \sigma^{\text{ref}} + (8\pi/15)^{1/2}\delta^i \sum_{m'} Y_{2,-m'}(\vartheta, \varphi) \right.$$

$$\left. \times \left[(3/2)^{1/2}\mathscr{D}^2_{0m'}(\Omega_i') + \tfrac{1}{2}\eta^i \left(\mathscr{D}^2_{2m'}(\Omega_i') + \mathscr{D}^2_{-2m'}(\Omega_i') \right) \right] \right\}. \tag{3-6}$$

ϑ and φ are external parameters. Six different, nondegenerate choices of ϑ, φ lead to six different line positions from which we may calculate, in principle, the desired parameters. This is, however, not the typical procedure.

Usually one rotates the crystal about an axis perpendicular to the external field and records what is called a rotation pattern. Stated differently, one records $\omega^i - \omega^{\text{ref}}$ when the external field moves in a plane of the crystal. Three rotation patterns with three different choices of the rotation axes are usually needed. The rotation patterns are always of the form

$$-[\omega^i(\Phi) - \omega^{\text{ref}}]/\omega_0 = C^i + A^i \cos 2(\Phi - \Phi^i_{\text{max}}), \tag{3-7}$$

where Φ is the rotation angle (counted from an arbitrary zero); C^i, A^i, and Φ^i_{max} are constants that are (complicated) functions of $\alpha_i', \beta_i', \gamma_i', \sigma^i - \sigma^{\text{ref}}, \delta^i, \eta^i$, and of the choice of the rotation axis. These functions become reasonably simple only if $\eta^i = 0$ (axially symmetric shielding tensor) and if we choose the axes $x_{\text{CR}}, y_{\text{CR}}, z_{\text{CR}}$ of the CRS as rotation axes. Figure 3-1 and Table 3-1 show how we get the desired internal parameters directly from the rotation patterns in such a simple case.

In the general case ($\eta \neq 0$) one usually determines these parameters by a least-squares computer fit analysis of the data of the rotation patterns. Equation (3-6) is a good starting point for the required program.

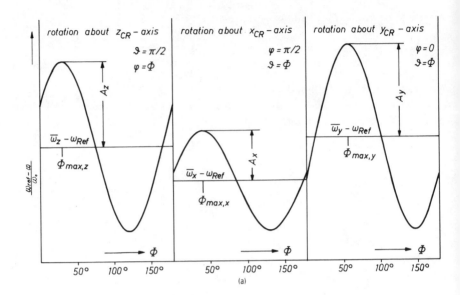

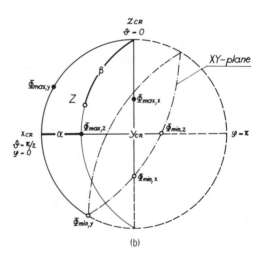

FIG. 3-1. (a) Example of rotation patterns when σ is axially symmetric; $\delta > 0$, $\alpha = 30°$, $\beta = 60°$ have been chosen. The lowest points in each pattern are the same. For $\delta < 0$ the highest points would be the same. Note what is plotted on the ordinate axis! ω is the resonance position of the line for $B_0 = $ const; $\omega_0 = \gamma_n B_0$.

(b) Stereographic projection of (1) the crystal fixed frame x_{CR}, y_{CR}, z_{CR}; (2) the unique shielding axis (Z) and the XY plane of the shielding principal axes system; (3) the paths of $\mathbf{B}_{st}$ when the crystal is rotated about the z_{CR}, y_{CR}, and x_{CR} axes; the rotation axis is always perpendicular to $\mathbf{B}_{st}$; (4) $\Phi_{max, x}$, $\Phi_{max, y}$, $\Phi_{max, z}$; $\Phi_{min, x}$, $\Phi_{min, y}$, $\Phi_{min, z}$.

When taking rotation patterns it is always wise to draw a figure like (b).

TABLE 3-1

CONSTANTS A, C, AND Φ_{max} IN EQ. (3-7) WHEN σ IS AXIALLY SYMMETRIC, AND RELATIONS FOR OBTAINING FROM THEM THE DESIRED PARAMETERS[a]

Rotation axis α	C_α	Φ_α	A_α	$\Phi_{max,\alpha}$
x_{CR}	$\sigma - \sigma_{ref} + \dfrac{\delta}{4}(1 - 3\sin^2\beta\cos^2\alpha)$	ϑ	$\dfrac{3}{4}\delta(1 - \sin^2\beta\cos^2\alpha)$	$-\dfrac{1}{2}\arcsin\left[\dfrac{\sin 2\beta \sin\alpha}{1 - \sin^2\beta\cos^2\alpha}\right]$
y_{CR}	$\sigma - \sigma_{ref} + \dfrac{\delta}{4}(1 - 3\sin^2\beta\sin^2\alpha)$	ϑ	$\dfrac{3}{4}\delta(1 - \sin^2\beta\sin^2\alpha)$	$-\dfrac{1}{2}\arcsin\left[\dfrac{\sin 2\beta \cos\alpha}{1 - \sin^2\beta\sin^2\alpha}\right]$
z_{CR}	$\sigma - \sigma_{ref} + \dfrac{\delta}{4}(1 - 3\cos^2\beta)$	φ	$\dfrac{3}{4}\delta\sin^2\beta$	α

$$\text{(a)} \quad \frac{1}{3}(C_x + C_y + C_z) = \sigma - \sigma_{ref}$$

$$\text{(b)} \quad A = A_x + A_y + A_z = \Delta\sigma$$

$$\text{(c)} \quad \sin^2\beta = 2\frac{A_z}{A}; \quad \sin 2\beta = \frac{2A_z \sin^2\Phi_{max,z} - A}{A \cos\Phi_{max,z}} \sin 2\Phi_{max,y}$$

$$\text{(d)} \quad \alpha = \Phi_{max,z}$$

[a] Equation (a) holds also in the general case ($\eta \neq 0$), as does Eq. (b) if at least one of the axes of the PAS coincides with one of the axes of the CRS, and provided $\Delta\sigma$ is interpreted as $\frac{1}{2}\delta(3+\eta) = \sigma_{zz} - \sigma_{xx}$. The sign of δ can be determined from whether the lowest or highest points in all patterns are the same or nearly the same for $1 \gg \eta \neq 0$. For $\delta < 0$ "max" must be replaced by "min."

C. Only Powders Available[10]

1. Powder Pattern Lineshapes

A powder consists of a multitude of randomly oriented single crystals. Consider a group of crystallographically equivalent nuclei and suppose that their shielding tensors are axially symmetric. If we plot the orientations of the unique axes of all these tensors in the powder sample on a sphere we get a distribution that is constant over the sphere. Alternatively, we may plot on a sphere the direction of $\mathbf{B}_{st}$ (specified by ϑ, φ) as seen in the PAS of each nucleus. This alternative makes it easy to drop the restricting assumption of axially symmetric shielding tensors.

The distribution we get is again constant over the sphere. Each point on the sphere corresponds to a definite orientation of $\mathbf{B}_{st}$ with respect to a shielding principal axes system and hence to a definite NMR spectral line position. It is obtained from Eq. (3-6) by setting $\Omega_i' = (0,0,0)$, which means making the CRS and the PAS coincide. Since $\mathscr{D}^2_{m''m'}(0) = \delta_{m''m'}$ we obtain

$$\omega^i - \omega^{\text{ref}} = -\omega_0\{\sigma^i - \sigma^{\text{ref}} + (8\pi/15)^{1/2}$$
$$\times \delta^i[(3/2)^{1/2}Y_{20}(\vartheta,\varphi) + \tfrac{1}{2}\eta^i(Y_{2,2}(\vartheta,\varphi) + Y_{2,-2}(\vartheta,\varphi))]\}$$
$$= -\omega_0\{\sigma^i - \sigma^{\text{ref}} + \delta^i[\tfrac{1}{2}(3\cos^2\vartheta - 1) + \tfrac{1}{2}\eta^i\sin^2\vartheta\cos 2\varphi]\}.$$

$$(3-8)$$

The orientational dependence of the line position is contained in

$$\omega = \omega_0\delta[\tfrac{1}{2}(3\cos^2\vartheta - 1) + \tfrac{1}{2}\eta\sin^2\vartheta\cos 2\varphi], \qquad (3-9)$$

where, for simplicity, we have dropped the indices i from the parameters η and δ. Using Eq. (3-9) we now can draw curves $\omega = $ const on our sphere. Two examples are given in Figs. 3-2 and 3-4.

The NMR spectrum of a powder sample is a superposition of the NMR lines from all the nuclei of all the grains of the sample. The intensity $I(\omega)$ of the spectrum, when integrated over an interval $\omega_a \cdots \omega_b$ is proportional to the number of nuclei whose NMR lines fall into that interval. This means it is proportional to the area between the curves $\omega = \omega_a$ and $\omega = \omega_b$ on our sphere (it is clear that we need consider only one hemisphere, in fact, only one octant of the sphere):

---

[10] The subject matter of this section is treated, e.g., by Bloembergen and Rowland.[11, 12] However, it is our experience that students often have a very hard time following their articles. Therefore, we discuss here the underlying ideas in a rather detailed manner and hope that this is warranted from a didactic point of view.

[11] N. Bloembergen and T. J. Rowland, *Acta Met.* **1**, 731 (1953).

[12] N. Bloembergen and T. J. Rowland, *Phys. Rev.* **55**, 1679 (1955).

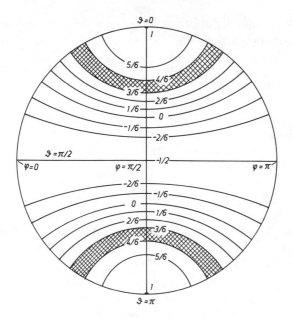

FIG. 3-2. Stereographic projection of curves $\omega = $ const (actually curves $\vartheta = $ const) when σ is axially symmetric, $\eta = 0$. Units are $\omega_0 \delta$. Notice the large distance of the curves for $\vartheta \to 0$ and $\vartheta \to \pi/2$.

$$\int_{\omega_a}^{\omega_b} I(\omega) \, d\omega = N \underset{\substack{\text{between} \\ \text{curves } \omega = \omega_a \\ \text{and } \omega = \omega_b}}{\iint} \sin \vartheta \, d\vartheta \, d\varphi. \qquad (3\text{-}10)$$

N is a normalization factor which we are free to choose in such a way that $\int_{-\infty}^{+\infty} I(\omega) \, d\omega = 1$. $N = 2/\pi$ if eventually we integrate over one octant of the sphere.

First let us consider the special case $\eta = 0$.

Figure 3-2 shows the curves $\omega = \omega_n = $ const. Equal increments of ω_n have been chosen. ω depends only on ϑ. The curves $\omega = $ const are actually curves $\vartheta = $ const. We integrate Eq. (3-10) over φ (from $\varphi = 0$ to $\varphi = \pi/2$). In the remaining integral over ϑ we introduce

$$\omega = \omega_0 \delta \frac{3 \cos^2 \vartheta - 1}{2}$$

as a new integration variable. This gives

$$\int_{\omega_a}^{\omega_b} I(\omega) \, d\omega = \frac{1}{\sqrt{3} \, \omega_0 \, \delta} \int_{\omega_a}^{\omega_b} \frac{d\omega}{[1 + (2\omega/\omega_0 \, \delta)]^{1/2}}. \qquad (3\text{-}11)$$

Equation (3-11) holds for arbitrary integration limits ω_a, ω_b; hence we can equate the integrands and obtain

$$I(\omega) = \frac{1}{\sqrt{3}\,\omega_0\,\delta}\,\frac{1}{[1+(2\omega/\omega_0\,\delta)]^{1/2}}\,,\qquad -\omega_0\,\delta/2 \leqslant \omega \leqslant \omega_0\,\delta.$$

$$(3\text{-}12)$$

Figure 3-3 shows $I(\omega)$, which is the idealized powder pattern when $\boldsymbol{\sigma}$ is axially symmetric. It diverges for $\omega \to -\tfrac{1}{2}\omega_0\,\delta$. We hope that Figure 3-2 provides a "look and see" understanding for this divergence. Note, in particular, the spacing of the curves $\omega = \omega_n$! Figure 3-3 also makes clear how the desired internal parameters $\sigma - \sigma^{\text{ref}}$ and $\Delta\sigma = \tfrac{3}{2}\delta$ can be determined from a powder pattern. Naturally no information about the direction of the unique axis relative to a crystal fixed frame can be derived from powder patterns.[13]

To treat the general case, $\eta \neq 0$, let us first introduce $x = -\cos\vartheta$ and then $\omega = \tfrac{1}{2}\omega_0\,\delta[3x^2 - 1 + \eta(1-x^2)\cos 2\varphi]$ as new integration variables in the right-hand integral of Eq. (3-10). The first step leads to

$$\frac{2}{\pi}\int\limits_{\cdots}^{\cdots}\!\!\int dx\,d\varphi,$$

where the dots indicate integration limits. The second step leads to

$$\int_{\omega_a}^{\omega_b} I(\omega)\,d\omega = \frac{2}{\pi}\int_{\omega_a}^{\omega_b}d\omega\int_{\varphi_l}^{\varphi_u}\Delta(\varphi,\omega)\,d\varphi.\qquad (3\text{-}13)$$

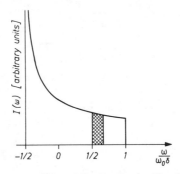

FIG. 3-3. Powder pattern when $\boldsymbol{\sigma}$ is axially symmetric. The divisions on the axis correspond to the lines $\omega = \text{const}$ in Fig. 3-2. Similarly, the shaded areas in Figs. 3-2 and 3-3 correspond to each other.

[13] By comparing ^{19}F multiple-pulse powder patterns of C_6F_6 at different temperatures and exploiting the knowledge that C_6F_6 molecule rotates rapidly about their 6-fold axes in the solid state at higher temperatures Mehring *et al.*[14] were able to determine the unique axis of the ^{19}F shielding tensor. C_6F_6, however, is an exceptional case.

[14] M. Mehring, R. G. Griffin, and J. S. Waugh, *J. Chem. Phys.* **55**, 746 (1971).

φ_l and φ_u are the lower and upper integration limits of the φ-integral, respectively. We shall specify them shortly. $\Delta(\varphi, \omega)$ is the Jacobian $\partial(\varphi, x)/\partial(\varphi, \omega)$, which reduces to

$$\frac{\partial x}{\partial \omega} = \frac{1}{\omega_3} \left[(3 - \eta \cos 2\varphi) \left(\frac{2\omega}{\omega_3} + 1 - \eta \cos 2\varphi \right) \right]^{-1/2}. \tag{3-14}$$

We have introduced the abbreviation $\omega_3 = \omega_0 \delta$ and similarly we shall define $\omega_2 = -\frac{1}{2}\omega_0 \delta(1-\eta) = \omega_0 \sigma_{YY}^{(2)}$ and $\omega_1 = -\frac{1}{2}\omega_0 \delta(1+\eta) = \omega_0 \sigma_{XX}^{(2)}$. By the same argument used in the paragraph following Eq. (3-11) we can now equate the integrands of the ω integrals in Eq. (3-13):

$$I(\omega) = \frac{2}{\pi \omega_3} \int_{\varphi_l}^{\varphi_u} (3 - \eta \cos 2\varphi)^{-1/2} \left(\frac{2\omega}{\omega_3} + 1 - \eta \cos 2\varphi \right)^{-1/2} d\varphi. \tag{3-15}$$

Let us now turn to φ_l and φ_u. We integrate only over one octant of the sphere; we choose the upper left one in Fig. 3-4. It is clear from Fig. 3-4 that we must distinguish two cases:

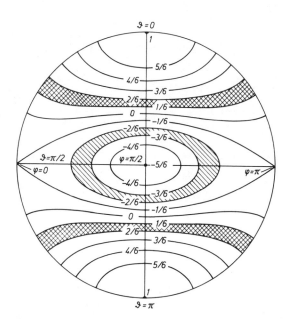

FIG. 3-4. Stereographic projection of curves $\omega = $ const when $\boldsymbol{\sigma}$ is nonaxially symmetric. $\eta = \frac{2}{3}$ has been chosen in this example. Units are $\omega_0 \delta = \omega_3$. For $\omega/\omega_3 < \omega_2/\omega_3 = -\frac{1}{2}(1-\eta)$ the curves cross the equator. The crossing meridian φ_l is determined by $\omega/\omega_3 = -\frac{1}{2}[1 - \eta \cos 2\varphi_l]$.

(i) $\quad \omega_2 \leqslant \omega \leqslant \omega_3 \to \varphi_l = 0, \ \varphi_u = \dfrac{\pi}{2}$,

(ii) $\quad \omega_1 \leqslant \omega \leqslant \omega_2 \to \varphi_l = \dfrac{1}{2}\arccos\left[\dfrac{1}{\eta}\left(1 + \dfrac{2\omega}{\omega_3}\right)\right], \ \varphi_u = \dfrac{\pi}{2}$.

To proceed, we introduce the new integration variable $z = \cos 2\varphi$. We get

$$I(\omega) = \frac{1}{\pi\omega_3\eta}\int_{z_l}^{z_u}\frac{dz}{[(1-z^2)(3/\eta-z)\{(1/\eta)[(2\omega/\omega_3)+1]-z\}]^{1/2}},$$

$$z_u = 1 \ \ (\text{case i}), \qquad z_u = \frac{1}{\eta}\left(\frac{2\omega}{\omega_3}+1\right) \ \ (\text{case ii}), \qquad z_l = -1 \ \ (\text{cases i, ii}).$$

This integral is tabulated, e.g., in the "Table of Integrals, Sums, Series and Products" by Gradshteyn and Ryzhik.[15] Note that the zeros z_n of the polynomial under the root are known; note also that the sequence of the z_n's (which is crucial) is different in cases i and ii.

Calculating the required parameters from z_l, z_u, and the z_n's we finally get

$$I(\omega) = \pi^{-1}[(\omega_3-\omega_2)(\omega-\omega_1)]^{-1/2}F\left\{\left(\frac{(\omega_3-\omega)(\omega_2-\omega_1)}{(\omega_3-\omega_2)(\omega-\omega_1)}\right)^{1/2}, \frac{\pi}{2}\right\}$$

$$= \pi^{-1}[(\omega_3-\omega_2)(\omega-\omega_1)]^{-1/2}K\left\{\arcsin\left[\frac{(\omega_3-\omega)(\omega_2-\omega_1)}{(\omega_3-\omega_2)(\omega-\omega_1)}\right]^{1/2}\right\}$$

$$\text{(3-16a)}$$

for $\omega_2 \leqslant \omega \leqslant \omega_3$, and

$$I(\omega) = \pi^{-1}[(\omega_3-\omega)(\omega_2-\omega_1)]^{-1/2}F\left\{\left[\frac{(\omega_3-\omega_2)(\omega-\omega_1)}{(\omega_3-\omega)(\omega_2-\omega_1)}\right]^{1/2}, \frac{\pi}{2}\right\}$$

$$= \pi^{-1}[(\omega_3-\omega)(\omega_2-\omega_1)]^{-1/2}K\left\{\arcsin\left[\frac{(\omega_3-\omega_2)(\omega-\omega_1)}{(\omega_3-\omega)(\omega_2-\omega_1)}\right]^{1/2}\right\}$$

$$\text{(3-16b)}$$

for $\omega_1 \leqslant \omega \leqslant \omega_2$.

$F(k, \varphi)$ is the (incomplete) elliptic integral of the first kind, and,

$$K(\arcsin k) = F(k, \pi/2)$$

is the complete elliptic integral of the first kind. We have normalized $I(\omega)$ such that the area under $I(\omega)$ is unity. Figure 3-5 shows an example. The discontinuities $\omega_3, \omega_2, \omega_1$ immediately yield $\sigma_{ZZ}, \sigma_{YY}, \sigma_{XX}$ or, alternatively, δ, η, and σ.

Before leaving the subject of nuclear-shielding powder lineshapes a final

---

[15] I. S. Gradshteyn and I. M. Ryzhik, "Table of Integrals, Sums, Series and Products," 4th ed., pp. 241–242. Academic Press, New York, 1966.

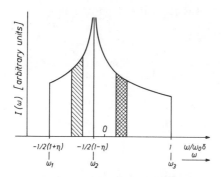

FIG. 3-5. Powder patterns when σ is nonaxially symmetric ($\eta = \frac{2}{3}$). The divisions on the axis correspond to the lines $\omega = $ const in Fig. 3-4. Again, the differently shaded areas in Figs. 3-4 and 3-5 correspond to each other. Two abscissa axes are indicated for convenience.

remark should be made. We have already emphasized that by investigating powders one does not get information about principal shielding directions. Hence single-crystal studies are always potentially superior to powder studies. In practice, however, there may be exceptions to this rule. Remember that all crystallographically equivlaent nuclei contribute to one and the same powder pattern. Now, in molecular solids the nuclear shielding is often dominated so much by molecular properties that not only all crystallographically, but all chemically equivalent nuclei (nuclei related by molecular symmetry operations) contribute to one and the same pattern. This may result for highly symmetric molecules (C_6F_6, P_4, C_6H_6, etc.) in powder patterns simple enough to be analyzed, whereas the corresponding single-crystal spectra may resist an analysis due to too large a number of often unresolved lines.

2. MOMENTS OF SHIELDING POWDER PATTERNS

The nth moment of a (normalized) lineshape function $I(\omega)$ or, in other terms, the mean value $\langle \omega^n \rangle$ of ω^n is defined as

$$\langle \omega^n \rangle = \int_{-\infty}^{+\infty} I(\omega) \, \omega^n \, d\omega. \tag{3-17}$$

This definition requires a specification of the point about which the moments are to be taken. We choose it such that $\langle \omega \rangle = 0$. This choice is consistent with Eq. (3-9). We call that point in the spectrum $\bar{\omega}$. $\bar{\omega}$ and $\langle \omega^2 \rangle$ of pure shielding powder patterns can be determined experimentally even when the experimental patterns are strongly affected by internal Hamiltonians other than $\mathscr{H}_{CS}$, e.g., by $\mathscr{H}_D$. The precise conditions under which this holds true are discussed by Van der Hart and Gutowsky.[16]

[16] D. L. Van der Hart and H. S. Gutowsky, *J. Chem. Phys.* **49**, 261 (1968).

A third quantity that is accessible experimentally is $\omega_{1/2}$ defined by

$$\int_{\omega_{1/2}}^{\infty} I(\omega)\, d\omega = \tfrac{1}{2}. \qquad (3\text{-}18)$$

$\omega_{1/2}$ divides the area under $I(\omega)$ in two equal halves. For symmetric patterns $\omega_{1/2} = \bar{\omega}$.

By inserting the squares of both sides of Eq. (3-9) in both sides of Eq. (3-10) and extending the integration limits appropriately, we obtain in a straight-forward manner

$$\langle \omega^2 \rangle = (\omega_0\, \delta)^2 \tfrac{1}{15}(3 + \eta^2). \qquad (3\text{-}19)$$

We were unsuccessful in finding an explicit formula for $\bar{\omega} = \omega_{1/2}$; therefore, we give in Fig. 3-6 a graphic plot of $f(\eta) = (\omega_0\, \delta)^{-1}|(\bar{\omega} - \omega_{1/2})|$ versus η. To determine δ and η (σ is fixed by $\bar{\omega}$ itself) from measured values of $\bar{\omega}$, $\omega_{1/2}$, and $\langle \omega^2 \rangle$ a plot of

$$g(\eta) = \langle \omega^2 \rangle^{-1/2}|\bar{\omega} - \omega_{1/2}| = f(\eta)\,[15/(3+\eta^2)]^{1/2}$$

is helpful and is given in Fig. 3-6. g is a measurable quantity. Knowing g we can immediately get η from Fig. 3-6. Knowing η and $\langle \omega^2 \rangle$ one can find δ from Eq. (3-19). The sign of δ is the same as the sign of $\bar{\omega} - \omega_{1/2}$.

This is a simple procedure to determine δ and η (and σ). However, we are afraid that it will rarely produce accurate results, at least as far as η is concerned. Its drawback is that $|\bar{\omega} - \omega_{1/2}|$ never exceeds $\tfrac{1}{12}$ of the total width of the spectrum. Therefore, nowadays most workers prefer to analyze their powder patterns by least-squares computer fit programs, as we do ourselves. Nevertheless, the procedure described above may help sometimes to get quickly results, if only preliminary ones.

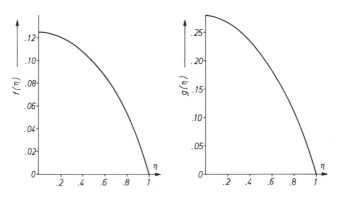

Fig. 3-6. The functions

$$f(\eta) = (\omega_0\delta)^{-1}|(\bar{\omega} - \omega_{1/2})| \qquad \text{and} \qquad g(\eta) = f(\eta)\,[15/(3+\eta^2)]^{1/2}.$$

D. Antisymmetric Constituents of σ

In the past few years there has been considerable discussion about a non-vanishing antisymmetric constituent of σ. The discussion runs, in fact, on two levels:

(i)   Do antisymmetric constituents exist at all?
(ii) Given they exist, how do they manifest themselves in NMR spectra?

The first of these questions has been answered in the affirmative, at least to the extent that there are no symmetry arguments forbidding an antisymmetric constituent of σ. Buckingham and Malm[17] have provided a valuable table that tells which of the elements of both the symmetric ($\sigma^{(s)}$) and antisymmetric ($\sigma^{(a)}$) constituents of σ can be nonzero for a given nuclear site symmetry. We shall draw upon that table later in this section. These authors have also described a model system with nonvanishing elements of $\sigma^{(a)}$. Schneider[18] and Griffin et al.[19] have dealt with the second of the above questions. They showed that $\sigma^{(a)}$ affects NMR line positions only in second order. We shall rederive their results here by a different, completely straightforward, and simple approach, and shall indicate how an experiment could be devised to demonstrate the existence of $\sigma^{(a)}$ in real physical systems.

In cartesian coordinates the Hamiltonian in question has the form

$$h\mathscr{H}_{CS}^{(a)} = \gamma_n \hbar \mathbf{I} \cdot \boldsymbol{\sigma}^{(a)} \cdot \mathbf{B}_{st}. \tag{3-20}$$

Again we drop indices designating nuclei. In the principal axes system of the *symmetric* constituent of σ we have

$$\boldsymbol{\sigma}^{(a)} = \begin{bmatrix} 0 & \sigma_{XY}^{(a)} & \sigma_{XZ}^{(a)} \\ -\sigma_{XY}^{(a)} & 0 & \sigma_{YZ}^{(a)} \\ -\sigma_{XZ}^{(a)} & -\sigma_{YZ}^{(a)} & 0 \end{bmatrix}. \tag{3-21}$$

We recall that in the axes system chosen the elements of $\sigma^{(a)}$ are identical with the off-diagonal elements of σ itself.

In irreducible spherical tensor notation $\mathscr{H}_{CS}^{(a)}$ reads as follows:

$$\mathscr{H}_{CS}^{(a)} = \gamma_n \sum_m (-1)^m R_{1,-m}^{CS} T_{1m}^{CS}, \tag{3-22}$$

[17] A. D. Buckingham and S. Malm, *Mol. Phys.* **22**, 1127 (1971).
[18] R. F. Schneider, *J. Chem. Phys.* **48**, 4905 (1968).
[19] R. G. Griffin, J. D. Ellett, M. Mehring, J. G. Bulitt, and J. S. Waugh, *J. Chem. Phys.* **57**, 2147 (1972); also see Kneubühl.[20]
[20] F. K. Kneubühl, *Phys. Kondens. Mater.* **1**, 410 (1963).

where

$$
\left.
\begin{aligned}
T_{11}^{\text{CS}} &= (1/\sqrt{2})(I_1 B_0 - I_0 B_1) \quad \rightarrow (1/\sqrt{2})I_1 B_0, \\
T_{10}^{\text{CS}} &= (1/\sqrt{2})(I_1 B_{-1} - I_{-1} B_1) \rightarrow 0, \\
T_{1,-1}^{\text{CS}} &= (1/\sqrt{2})(I_0 B_{-1} - I_{-1} B_0) \rightarrow -(1/\sqrt{2})I_{-1} B_0.
\end{aligned}
\right\}
\tag{3-23}
$$

The expressions following the arrows in Eqs. 3-23 give the T_{1m}^{CS}'s for our usual choice of $\mathbf{B}_{\text{st}}$. Again, we denote the R_{1m}^{CS}'s in the principal axes system of $\boldsymbol{\sigma}^{(s)}$ by ρ_{1m}^{CS}. The ρ_{1m}^{CS} are given in terms of the cartesian components of $\boldsymbol{\sigma}^{(a)}$ by

$$
\rho_{11}^{\text{CS}} = (\sigma_{XZ}^{(a)} + i\sigma_{YZ}^{(a)}),
\tag{3-24a}
$$

$$
\rho_{10}^{\text{CS}} = -i\sqrt{2}\,\sigma_{XY}^{(a)},
\tag{3-24b}
$$

$$
\rho_{1,-1}^{\text{CS}} = (\sigma_{XZ}^{(a)} - i\sigma_{YZ}^{(a)}).
\tag{3-24c}
$$

Equation (2-20), with $l = 1$ expresses the R_{1m}^{CS}'s needed in Eq. (3-22) by the ρ_{1m}^{CS}'s and by the elements of the Wigner rotation matrix $\mathscr{D}_{mm'}^1$:

$$
R_{1m}^{\text{CS}} = \sum_{m'} \mathscr{D}_{m'm}^1(\Omega)\,\rho_{1m'}^{\text{CS}}.
\tag{3-25}
$$

The Euler angles $(\alpha, \beta, \gamma) \equiv \Omega$ are the same as, e.g., in Eq. (2-28). This is assured by our choice of the frame of reference in which we specified the ρ_{1m}^{CS}'s.

By inserting Eq. (3-25) and Eqs. (3-23a)–(3-23c) into Eq. (3-22) we get $\mathscr{H}_{\text{CS}}^{(a)}$ in a form that is convenient for our discussion:

$$
\begin{aligned}
\hbar\mathscr{H}_{\text{CS}}^{(a)} &= \hbar\gamma_{\text{n}} \sum_{mm'} (-1)^m \mathscr{D}_{m',-m}^1 \rho_{1m'}^{\text{CS}}\, T_{1m}^{\text{CS}} \\
&= \hbar\omega_0 \frac{1}{\sqrt{2}} \sum_{m'} (\mathscr{D}_{m'1}^1 I_{-1} - \mathscr{D}_{m',-1}^1 I_1)\,\rho_{1m'}^{\text{CS}}.
\end{aligned}
\tag{3-26}
$$

We approach the manifestation of $\mathscr{H}_{\text{CS}}^{(a)}$ in NMR spectra of solids by perturbation theory. The total Hamiltonian that we consider is

$$
\hbar\mathscr{H} = \hbar\mathscr{H}_{\text{Z}} + \hbar\mathscr{H}_{\text{CS}}^{(s)} + \hbar\mathscr{H}_{\text{CS}}^{(a)},
\tag{3-27}
$$

where $\hbar\mathscr{H}_{\text{Z}}$ is the unperturbed Hamiltonian. The perturbation approach will work well so long as all $\sigma_{\alpha\beta} \ll 1$.

For $I = \tfrac{1}{2}$, $\mathscr{H}_{\text{Z}}$ defines two eigenstates, $|-\rangle$ and $|+\rangle$, with energies $\tfrac{1}{2}\hbar\gamma_{\text{n}} B_0 = \tfrac{1}{2}\hbar\omega_0$, and $-\tfrac{1}{2}\hbar\omega_0$. In first order only the so-called secular terms $\mathscr{H}_{\text{CS, secular}}^{(s)}$ of $\mathscr{H}_{\text{CS}}^{(s)}$ [see Eq. (2-29)] cause energy level shifts. The corresponding resonance shifts in the NMR spectrum have been discussed in detail in Section B. $\mathscr{H}_{\text{CS}}^{(a)}$, in particular, has no diagonal elements between the eigenstates of $\mathscr{H}_{\text{Z}}$. This is immediately clear from Eq. (3-26). Both $\mathscr{H}_{\text{CS}}^{(a)}$ and the nonsecular terms of $\mathscr{H}_{\text{CS}}^{(s)}$ do cause second-order energy shifts. For the upper level ($|-\rangle$) the shift is

$$\varepsilon_u = |\langle - | \, \hbar \mathscr{H}_{CS}^{(a)} + \hbar \mathscr{H}_{CS,\,\text{nonsecular}}^{(s)} \, | + \rangle|^2 / \hbar \omega_0 ;$$

for the lower level ($|+\rangle$) it is $\varepsilon_l = -\varepsilon_u$. Hence, the second-order angular frequency shift is

$$\Delta \omega_2 = 2\varepsilon_u / \hbar$$

$$= \omega_0 \left| \langle - | \left\{ \sum_{m'} \mathscr{D}_{m'1}^1 \, \rho_{1m'}^{CS} - \delta \left[\sqrt{\tfrac{3}{2}} \, \mathscr{D}_{01}^2 + \tfrac{1}{2}\eta \, (\mathscr{D}_{2,1}^2 + \mathscr{D}_{-2,1}^2) \right] \right\} I_{-1} | + \rangle \right|^2$$

$$= \tfrac{1}{2}\omega_0 |\{\cdots\}|^2 . \tag{3-28}$$

Note that the index m is one in all the $\mathscr{D}_{m'm}^l$ that enter Eq. (3-28). This means that every term has a factor $e^{i\alpha}$ (see Table 2-1), which drops out on performing the $|\cdots|^2$ operation: rotations of the crystal about the magnetic field are unobservable by NMR.

Equation (3-28) tells us that the resonance position of our spectral line is affected by the antisymmetric constituent of σ only in second order. However, the second-order shift $\Delta \omega_2$ is not entirely due to $\sigma^{(a)}$; a considerable part of it arises from $\sigma^{(s)}$. We now discuss a specific example.

We shall not consider a completely general antisymmetric tensor but one whose XY (and YX) components only are nonzero. From the spherical components of $\sigma^{(a)}$ only ρ_{10}^{CS} will be nonzero as a consequence. According to Buckingham and Malm's table[17] this choice of $\sigma^{(a)}$ applies for C_2, C_s, and C_{2h} nuclear site symmetries. Equation (3-28) simplifies in this special case:

$$\Delta \omega_2 = \tfrac{1}{2}\omega_0 \left| \mathscr{D}_{01}^1 \, \rho_{10}^{CS} - \delta \left[\sqrt{\tfrac{3}{2}} \, \mathscr{D}_{01}^2 + \tfrac{1}{2}\eta \, (\mathscr{D}_{2,1}^2 + \mathscr{D}_{-2,1}^2) \right] \right|^2$$

$$= \tfrac{1}{2}\omega_0 \left| i \sin \beta \, \sigma_{XY}^{(a)} - \delta \left[-\tfrac{3}{4} \sin 2\beta \right. \right.$$

$$\left. \left. + \tfrac{1}{4}\eta \sin \beta \, ((1 + \cos \beta) \, e^{2i\gamma} - (1 - \cos \beta) \, e^{-2i\gamma}) \right] \right|^2$$

$$= \tfrac{1}{2}\omega_0 \{ \sin^2 \beta \, (\sigma_{XY}^{(a)})^2 - \delta \eta \sigma_{XY}^{(a)} \sin^2 \beta \, \sin 2\gamma + \tfrac{1}{4}\delta^2 \eta^2 \sin^2 \beta \, \sin^2 2\gamma$$

$$+ \tfrac{1}{16}\delta^2 \sin^2 2\beta \, (3 - 6\eta \cos 2\gamma + \eta^2 \cos^2 2\gamma) \} . \tag{3-29}$$

We now ask ourselves how does the second-order shift manifest itself in rotation patterns. In the spirit of our previous discussion of rotation patterns we rotate the magnetic field $\mathbf{B}_{st}$ about three orthogonal crystal fixed axes. For simplicity we choose the principal axes system of $\sigma^{(s)}$ of the nucleus considered as crystal fixed axes system. Thus we rotate $\mathbf{B}_{st}$ successively in the XY, XZ, and YZ planes. For ease of comparison we also indicate the first-order shifts $\Delta \omega_1$. These are obtained in a straightforward way from Eq. (2-30).

(i) $\mathbf{B}_{st}$ rotates in the XY plane (γ is running, $\beta = \tfrac{1}{2}\pi$):

$$\Delta\omega_2\,(\gamma;\,\beta=\tfrac{1}{2}\pi)\;=\;\tfrac{1}{2}\omega_0\,\{(\sigma_{XY}^{(a)})^2-\delta\eta\sigma_{XY}^{(a)}\sin 2\gamma+\tfrac{1}{4}\delta^2\eta^2\sin^2 2\gamma\}$$

$$=\tfrac{1}{2}\omega_0\,\{[(\sigma_{XY}^{(a)})^2+\tfrac{1}{8}\delta^2\eta^2]-\delta\eta\sigma_{XY}^{(a)}\sin 2\gamma-\tfrac{1}{8}\delta^2\eta^2\cos 4\gamma\},$$

$$(3\text{-}30\mathrm{a})$$

$$\Delta\omega_1\,(\gamma;\,\beta=\tfrac{1}{2}\pi)\;=\;-\omega_0\{[\sigma-\sigma^{\mathrm{ref}}-\tfrac{1}{3}\delta]+\tfrac{1}{2}\delta\eta\cos 2\gamma\}. \qquad (3\text{-}30\mathrm{b})$$

(ii) $\mathbf{B}_{\mathrm{st}}$ rotates in the XZ plane (β is running, $\gamma=0$):

$$\Delta\omega_2\,(\beta;\,\gamma=0)\;=\;\tfrac{1}{2}\omega_0\,\{\sin^2\beta(\sigma_{XY}^{(a)})^2+\tfrac{1}{16}(3-\eta)^2\sin^2 2\beta\} \qquad (3\text{-}31\mathrm{a})$$

$$=\tfrac{1}{2}\omega_0\,\{[\tfrac{1}{2}(\sigma_{XY}^{(a)})^2+\tfrac{1}{32}\delta^2(3-\eta)^2]$$

$$-\tfrac{1}{2}(\sigma_{XY}^{(a)})^2\cos 2\beta-\tfrac{1}{32}\delta^2(3-\eta)^2\cos 4\beta\},$$

$$\Delta\omega_1\,(\beta;\,\gamma=0)\;=\;-\omega_0\{[\sigma-\sigma^{\mathrm{ref}}+\tfrac{1}{4}\delta(1+\eta)]+\tfrac{1}{4}\delta(3-\eta)\cos 2\beta\}.$$

$$(3\text{-}31\mathrm{b})$$

(iii) $\mathbf{B}_{\mathrm{st}}$ rotates in the YZ plane (β is running, $\gamma=\tfrac{1}{2}\pi$)

$$\Delta\omega_2\,(\beta;\,\gamma=\tfrac{1}{2}\pi)\;=\;\tfrac{1}{2}\omega_0\,\{\sin^2\beta(\sigma_{XY}^{(a)})^2+\tfrac{1}{16}\delta^2(3+\eta)^2\sin^2 2\beta\}$$

$$=\tfrac{1}{2}\omega_0\,\{[\tfrac{1}{2}(\sigma_{XY}^{(a)})^2+\tfrac{1}{32}\delta^2(3+\eta)^2]$$

$$-\tfrac{1}{2}(\sigma_{XY}^{(a)})^2\cos 2\beta-\tfrac{1}{32}\delta^2(3+\eta)^2\cos 4\beta\}, \qquad (3\text{-}32\mathrm{a})$$

$$\Delta\omega_1\,(\beta;\,\gamma=\tfrac{1}{2}\pi)\;=\;-\omega_0\,\{[\sigma-\sigma^{\mathrm{ref}}+\tfrac{1}{4}\delta(1-\eta)]+\tfrac{1}{4}\delta(3+\eta)\cos 2\beta\}.$$

$$(3\text{-}32\mathrm{b})$$

Equations (3-30)–(3-32) show that the inclusion of second-order terms—quadratic in the shielding components—results in a number of new features of the rotation patterns:

(a)   The rotation patterns are no longer purely π periodic. They do have terms that are $\tfrac{1}{2}\pi$ periodic ($\cos 4\gamma$; $\cos 4\beta$). This has been emphasized by Schneider.[18] However, the $\tfrac{1}{2}\pi$-periodic terms are not related to antisymmetric shielding components. Hence $\tfrac{1}{2}\pi$-periodic terms in rotation patterns are not manifestations of $\boldsymbol{\sigma}^{(a)}$. They should alert the experimenter to look for manifestations of $\boldsymbol{\sigma}^{(a)}$ in the π-periodic terms but more so to check his experimental setup, because most probably they are an indication that something went wrong in the experiment.

(b)   The average shift is no longer $\sigma-\sigma^{\mathrm{ref}}$. However, by considering the average shift at just one field strength B_0 there is no way to detect the difference.

(c)   For our special choice of $\boldsymbol{\sigma}^{(a)}$ the most conspicuous manifestation of antisymmetric shielding occurs in the XY-rotation pattern where $\sigma_{XY}^{(a)}$ enters linearly. In the XZ- and YZ patterns $\boldsymbol{\sigma}^{(a)}$ enters squared and its effect is only to simulate slightly modified values of the anisotropy and asymmetry parameters.

Let us focus our attention on two nuclei, 1 and 2, related by a center of symmetry. The space group $Pcmn$, e.g., is consistent with the symmetry requirements for Eqs. (3-30a)–(3-32b) and the presence of a center of symmetry. For $\sigma^{(a)} \equiv 0$ the NMR lines from nuclei 1 and 2 coincide. For $\sigma^{(a)} \neq 0$ they are split in the XY-rotation pattern. By definition the sign of $\sigma_{XY}^{(a)}$ is opposite for two nuclei related by a center of symmetry.

Equation (3-30a) tells us how large, or rather how small, the splitting $\delta\omega$ will be:

$$|\delta\omega/\omega_0| = |\delta\eta\sigma_{XY}^{(a)} \sin 2\gamma|. \qquad (3\text{-}33a)$$

The maximum splitting will be reached for $\sin 2\gamma = 1$:

$$|\delta\omega_{max}/\omega_0| = |\delta\eta\sigma_{XY}^{(a)}|. \qquad (3\text{-}33b)$$

Naturally the splitting is quadratically small in shielding parameters $(\delta, \sigma_{XY}^{(a)})$. Unfortunately, $\eta \leqslant 1$ enters as an additional factor.

For light nuclei such as ^{1}H and ^{13}C, where δ and—hopefully—$\sigma_{XY}^{(a)}$ may be of the order of 2×10^{-5} and 2×10^{-4}, respectively, there is nowadays hardly a chance to observe such a line splitting due to $\sigma^{(a)}$. We repeat: This line splitting is the most conspicuous manifestation of $\sigma^{(a)}$. On the other hand, for heavy nuclei such as ^{207}Pb, where δ and $\sigma_{XY}^{(a)}$ may range up to 10^{-2} or even higher, an experimental demonstration of the line splitting described above and hence a detection of $\sigma^{(a)}$ seems feasible under favorable circumstances.

We warned above against experimental pitfalls. One that an experimenter must be prepared to encounter and that is liable to produce "false" line splitting is small-angle twinning of the sample crystal. Equal intensities of the two components of the split line, or even better unsplit resonance lines from a pair of quadrupolar nuclei related by a center of symmetry, are valuable and convincing checks for the absence of crystal twinning.

IV. Averaging in Ordinary Coordinate and Spin Spaces

A. Introductory Remarks

Our experimental access to NMR parameters rests largely on averaging in both ordinary and spin spaces. Some of the internal Hamiltonians or, occasionally, certain parts of them are effectively eliminated by these averaging processes. As a result others are left with fewer competitors. Part of this averaging occurs quite naturally, e.g., in liquid samples. An example of natural averaging in spin space is the famous truncation of internal Hamiltonians appropriate in high applied fields $\mathbf{B}_{st}$. Talking about averaging in these cases is merely one of several equivalent ways of describing a natural physical situation. There are, however, other cases where experimenters have taken active steps to remove the effects on NMR spectra of some of the physically present internal Hamiltonians in order to make the parameters of others more readily measurable. Their way of thinking was and still is in terms of averaging. This is so fruitful because it enables them to judge the result of a proposed procedure such as a new multiple-pulse sequence from an appropriate effective Hamiltonian and relieves them of the laborious task of solving the Schrödinger equation in each particular case.

The averages in question are time averages. Our first task is obviously to render our internal Hamiltonians time dependent. As we introduced them they are time independent for two reasons. First, we assumed tacitly that the parameters, essentially the R_{lm}'s, are time independent. Second, we used the Schrödinger representation in which the operators, essentially the T_{lm}'s, are time independent and in which the time dependence of the quantum mechanical system (the spin system) is contained entirely in the states or,

37

alternatively, in the density matrix ρ. Consequently there are two ways to render the internal Hamiltonians time dependent.

First, the parameters can be made time dependent. This introduces an explicit time dependence into the Hamiltonian. The way to do it is to impose motions on the molecules or crystals to which the parameters are linked. Examples are melting the sample and spinning the sample about axes making certain special angles with $\mathbf{B}_{st}$.

Second, the operators can be made time dependent. This can be accomplished by using interaction representations. Of course, the representation we use cannot and does not have any effect upon the *spectra*. However, by using an appropriate interaction representation we can treat *both* ways of introducing time dependences into our Hamiltonians in the *same* manner and make use of an important general rule (to be proven in Section D), which states that spectra are affected in lowest order only by the time-independent parts or time averages of the internal Hamiltonians. (These parts will also be called average or effective Hamiltonians.) The lowest order description of the spectra will be the better the higher the frequency scale of the time dependence involved is compared with the lineshifts or splittings the same Hamiltonians would produce in the spectra in the absence of the time dependence. In Section D we shall also deal with the problem of what corrections should be applied when this lowest order description no longer is adequate.

A crucial question is now, of course, which is the appropriate interaction representation. The general answer is: the one that puts *common*, in particular fast common, motions of the spins (or groups of spins) on the operators and leaves relative motions on the states (or on ρ). Common sense is called for in applying this rule. There will be both ambiguous and clear-cut cases. Only in the clear-cut cases is it wise to draw upon the general rule stated above.

The terms of the total Hamiltonians that we shall remove by using interaction representations will always be single-spin Hamiltonians. The interaction representations will thus be equivalent to viewing the spins from moving spin frames of reference. Often the space and spin frames of reference we shall use will be different and will be accelerated with respect to each other. One consequence of this schizophrenic situation is that Hamiltonians and density matrices expressed in these frames will not be scalars in the usual sense of the term.[6]

We shall now consider the averaging consequences of a number of motions in ordinary coordinate space. Then we shall turn to motions in spin space. Only averages will be taken into account in this section. These are the lowest order approximations in each particular case. Experimenters strive to create situations where this lowest order approximation provides an adequate description of the system. Corrections are discussed in Sections D–F and in Chapter V.

B. Averaging in Ordinary Coordinate Space

The following types of motions are of greatest importance in NMR spectroscopy:

(1) Random, isotropic translational and reorientational motions of molecules.

(2) Random translational and anisotropic reorientational motions of molecules.

(3) Random jumps of molecules or mobile molecular parts between two or more different sites, or between two or more different isomeric conformations.

(4) Sample spinning about an axis perpendicular to the applied field.

(5) Sample spinning about an axis at an angle $\beta_m = \arccos(1/\sqrt{3}) \approx 54°44'$ to the applied field $\mathbf{B}_{st}$.

This list, of course, by no means exhausts the motions that are important in NMR. The first three types of motions in our list occur on a molecular scale and are, in a sense, provided by nature. The last two are macroscopic motions of the bulk sample and are deliberately implemented by the experimenter. We shall now discuss how these motions affect the measurability of NMR parameters.

1. RANDOM, ISOTROPIC TRANSLATIONAL AND REORIENTATIONAL MOTIONS OF MOLECULES

Such motions are characteristic of gases, isotropic liquids, and isotropic solutions. All R_{lm}, $l > 0$, vanish when averaged over such motions. Only the scalar ($l = 0$) chemical shifts and molecular scalar indirect spin–spin couplings survive. These are the interactions that cause the structures of high-resolution NMR spectra and from such spectra we can obtain information only about the corresponding parameters. We stress that (scalar) indirect spin–spin couplings between spins on different molecules are averaged out by the random translational components of the motions. Intermolecular magnetic-shielding effects give rise to so called "solvent effects."

2. RANDOM, ISOTROPIC TRANSLATIONAL AND ANISOTROPIC REORIENTATIONAL MOTIONS OF MOLECULES

Such motions are characteristic of, in particular, nematic liquids and solutions. Unrestricted translational motions ensure in the long run that all *inter*molecular interactions are wiped out. *Intra*molecular isotropic as well as anisotropic interactions survive, the latter only in part. The NMR lines arising from the single-spin Hamiltonians $\mathscr{H}_{CS}$ and $\mathscr{H}_Q$ are split by the spin coupling Hamiltonians $\mathscr{H}_D$ and $\mathscr{H}_J$. As the spin couplings are restricted to spins on one and the same molecule, only a finite number of spin couplings exist. As a result the spectra consist of a finite, though typically fairly large, number of often resolved lines. Such spectra, together with the high-resolution spectra of the same liquids or solutions in their isotropic phases enable one to obtain

(a)   the scalar chemical shifts σ^i and the scalar spin–spin coupling constants $J^{i,k}$, and

(b)   one relation for each of the tensor parameters $(\mathbf{D}+\mathbf{J}^{(2)})^{i,k}$, $\mathbf{V}^i$, and $\sigma^{(2),i}$, provided the so-called orientation matrix[21] S_{ij} is known. Antisymmetric constituents of σ and $\mathbf{J}$—if they exist at all—vanish because S_{ij} is symmetric.

Only a few experiments have been done on nuclei with $I > \frac{1}{2}$ and hence $\mathscr{H}^Q \neq 0$. $\mathbf{J}^{(2)i,k}$ is usually neglected in the first steps of the interpretation of the spectra and the results are used to get information about the geometry of the molecule (via $\mathbf{D}^{i,k}$). Complete knowledge about the shape of a molecule can be obtained only if the resonances from a sufficiently large number of nuclei are observed. It turns out that from nematic-solution NMR spectra alone, only ratios of internuclear distances, but never absolute distances, can be determined.[22]

Note that relations exist among the intermolecular distances r_{ik} for molecules with more than four atoms. The reader may convince himself of this fact by trying to attach a fifth atom to a four-atom molecule: he is free to choose arbitrary values for, e.g., r_{15}, r_{25}, and r_{35}, but r_{45} will then be fixed. This is so for purely geometric reasons. For molecules that possess symmetries, such relations exist even if the molecules have only four or less atoms. Experimental r_{ik}'s in the molecule CH_3F, determined neglecting $\mathbf{J}^{(2)i,k}$, turned out to be inconsistent with such relations. This has been interpreted by Krugh and Bernheim[23] as evidence for nonnegligible elements of $\mathbf{J}^{(2)F,H}$. This interpretation has been contested, however, by Bulthuis and MacLean[24]. More recent work along the same lines allowed reliable measurements of $\mathbf{J}$-coupling anisotropies[25,26] and, at least in one case (1,1-difluoroethene), the determination of a full F—F J-coupling tensor.[27]

[21] A. Saupe, Z. Naturforsch. A **19**, 161 (1964).
[22] P. Diehl and C. L. Khetrapal, "Basic Principles and Progress in NMR," Vol. 1. Springer-Verlag, Berlin and New York, 1969.
[23] T. R. Krugh and R. A. Bernheim, J. Chem. Phys. **52**, 4942 (1970).
[24] J. Bulthuis and C. MacLean, J. Magn. Resonance **4**, 148 (1971).
[25] J. Gerritsen, G. Koopmans, H. S. Rollema, and C. MacLean, J. Magn. Resonance **8**, 20 (1972).
[26] G. J. Den Otter, J. Gerritsen, and C. MacLean, J. Mol. Struct. **16**, 379 (1973).
[27] J. Gerritsen and C. MacLean, J. Magn. Resonance **5**, 44 (1971).

The one relation obtained for the second-rank constituent $\sigma^{(2)}$ of the shielding tensor provides complete information if σ is known to be axially symmetric by molecular symmetry. The unique axis is then, of course, also fixed by molecular symmetry. Sometimes, where such information was not available it was replaced tacitly or explicitly by a corresponding assumption.[28,29] However, it turned out that such assumptions are very dangerous.[14] Nevertheless, until the advent of multiple-pulse and spin-decoupling techniques applicable to solids and the availability of very high magnetic fields the bulk of our knowledge about nuclear magnetic-shielding anisotropies was derived from NMR spectra of nematic solutions. Excellent reviews of this field have recently been provided by Diehl and Khetrapal,[22] and by Appleman and Dailey.[30]

We conclude this subsection by pointing out that—as far as information is concerned—a nematic-solvent NMR experiment is equivalent to an NMR experiment on a single crystal with magnetically equivalent, magnetically well-isolated molecules. In this analogy, however, only one orientation of the single crystal is available since the ordering of the nematic solvent occurs in a fixed orientation with respect to the applied field. Recently it has been shown that in some favorable cases it is possible to change the average orientation of solute molecules with respect to the nematic solution by changing the temperature of the sample. An example is 1,2-difluoroethene, where this technique enabled Gerritsen and MacLean[27] to measure the J^{FF} tensor, and to test the assumption that the shielding of the ^{19}F nucleus is axially symmetric about the C—F bond in this molecule.[31]

3. RANDOM MOLECULAR JUMPS

Such motions are encountered in both fluids and solids. The only interactions not averaged out in fluids by isotropic random translational and reorientational motions are scalar chemical shifts and scalar spin–spin interactions. They can still be affected by random jumps of mobile molecular parts between two or more different sites, or between two or more different isomeric conformations. One speaks, in particular, of "chemical exchange" when the motion carries back and forth a spin between two or more environments in which it experiences different isotropic chemical shifts. If the motion is fast enough one sees only one line at the position of the weighted average of the different site chemical shifts. The weights are the probabilities of the spin being in the particular sites. One example that has even found its way into

[28] L. Snyder, J. Chem. Phys. 43, 4041 (1965).
[29] J. Biemond, B. J. M. Neyzen, and C. MacLean, Chem. Phys. 1, 335 (1973).
[30] B. R. Appleman and B. P. Dailey, Advan. Magn. Resonance 7, 231 (1974).
[31] G. J. Den Otter and C. MacLean, Chem. Phys. Lett. 20, 306 (1973).

Scientific American is cyclohexane.[32] Cyclohexane molecules "jump" between two different but completely equivalent (chair) conformations, exchanging thereby their axial and equatorial protons. At $T < -60°C$ the exchange is "slow" and two well-separated lines are observed. At room temperature the exchange is "fast" and only a single line is seen. The transition with rising temperature from a two- to a single-line spectrum provides an excellent means of studying the exchange rate.

The spectral changes from "slow" to "fast" exchange have been treated by Anderson[33] and others.[34,35] The theory has recently been extended to molecular reorientational jumps and transitions between different isomeric conformations in solids.[36] Powder patterns as described for rigid solids in Chapter III are affected in a characteristic way: in the region of the intermediate exchange rate they display dips instead of varying monotonically between the discontinuities of the (idealized) "rigid" powder patterns. Again, studies of spectral changes with temperature yield detailed information about the rate and also type of molecular motions.[37]

4. SAMPLE SPINNING ABOUT AN AXIS PERPENDICULAR TO THE APPLIED FIELD

This motion does not affect, of course, scalar parameters. It does affect tensor parameters. As we shall see, however, second-rank tensors are not wiped out completely. The realm of this kind of sample spinning is high-resolution NMR in fluids, where the tensor parameters have been wiped out already by random molecular motions. Its prime purpose is to reduce the inhomogeneity broadening of inherently very narrow NMR lines. We discuss this scheme of enhancement of resolution briefly here. Its effect upon internal Hamiltonians is considered in the following subsection.

The field strength at $\mathbf{r} = (r, \Theta, \Phi) \cong (x, y, z)$, and consequently the resonance position of a line contribution originating from a volume element at $\mathbf{r}$, can be described in a completely general way by

$$B_0(\mathbf{r}) = B_0(0) + \sum_{l=1}^{\infty} \left(\frac{4\pi}{2l+1}\right)^{1/2} \sum_{m=-l}^{+l} a_{lm} r^l Y_{lm}(\Theta, \Phi). \qquad (4\text{-}1)$$

$a_{l,-m} = (-1)^m a_{lm}^*$, since Eq. (4-1) describes a real field.

[32] *Scientific American*, **222**, 65 (1970).
[33] P. W. Anderson and P. R. Weiss, *J. Phys. Soc. Jap.* **9**, 316 (1954).
[34] S. Alexander, A. Baram, and Z. Luz, *Mol. Phys.* **27**, 441 (1974).
[35] H. Sillescu, *J. Chem. Phys.* **54**, 2110 (1971).
[36] H. W. Spiess, *Chem. Phys.* **6**, 217 (1974).
[37] H. W. Spiess, R. Grosescu, and U. Haeberlen, *Chem. Phys.* **6**, 226 (1974).

$B_0(\mathbf{r}) - B_0(0)$

$$= \sum_{l=1}^{\infty} \left(\frac{4\pi}{2l+1}\right)^{1/2} r^l \left\{ a_{l0} Y_{l0}(\Theta,\Phi) + \sum_{m=1}^{l} \left(a_{lm} Y_{lm}(\Theta,\Phi) + a_{lm}^* Y_{lm}^*(\Theta,\Phi)\right) \right\}$$

$$= a_{10} z - \mathrm{Re}\, a_{11} \sqrt{2}\, x + \mathrm{Im}\, a_{11} \sqrt{2}\, y \qquad (l=1)$$

$$+ a_{20} \frac{3z^2 - r^2}{2} + \mathrm{Re}\, a_{22} \sqrt{\frac{3}{2}}(x^2 - y^2)$$

$$+ \sqrt{6}\,(\mathrm{Im}\, a_{21}\, yz - \mathrm{Re}\, a_{21}\, xz - \mathrm{Im}\, a_{22}\, xy) \qquad (l=2)$$

$$+ \cdots. \tag{4-2}$$

Users of high-resolution spectrometers will recognize in the polynomials of Eq. (4-2) the names of the knobs of their shim control units. Turning up these knobs actually means "turning" up the respective coefficients $\mathrm{Re}\, a_{lm}$ or $\mathrm{Im}\, a_{lm}$.

When the sample is spun all its volume elements (except those on the rotation axis) change their positions with respect to a space fixed frame and, hence, sweep through a sequence of different field values. The field a specific volume element with coordinates r, Θ, Φ in a sample fixed frame sees during rotation of the sample is given by

$$B_0(\mathbf{r}(t)) = B_0(0) + \sum_{l=1}^{\infty} \left(\frac{4\pi}{2l+1}\right)^{1/2} r^l \sum_{m=-l}^{+l} a_{lm} \sum_{m'=-l}^{+l} \mathscr{D}_{m'm}^l(0,\omega_r t,0) Y_{lm'}(\Theta,\Phi),$$

if we specifically rotate the sample about the y-axis of the space fixed frame with angular velocity ω_r. As we stated in the introduction of this chapter, all we need to retain is $\overline{B_0(\mathbf{r}(t))}^t$ or, as a consequence, $\overline{\mathscr{D}_{m'm}^l(0,\omega_r t,0)}^t$. With the aid of Table 2-1 (and a corresponding table of the $\mathscr{D}_{m'm}^1$, which can be found, e.g., in Edmonds[1]) we get up to and including terms with $l = 2$:

$$\overline{B_0(\mathbf{r}(t))}^t = B_0(0) + \mathrm{Im}\, a_{11} \sqrt{2}\, y + a_{20}\{\tfrac{1}{8}[3z^2 - r^2 + 3(x^2 - y^2)]\}$$

$$+ \mathrm{Re}\, a_{22} \tfrac{1}{4}\sqrt{\tfrac{3}{2}}(3z^2 - r^2 + x^2 - y^2) + \cdots. \tag{4-3}$$

It is clear from Eq. (4-3) that the effective field variation over the sample is reduced substantially by sample spinning. However, a number of terms in the expansion survive. Manufacturers of high-homogeneity magnets pay special attention to minimizing the corresponding coefficients.

5. MAGIC-ANGLE SAMPLE SPINNING

With magic-angle sample spinning we are entering the realm of high-resolution NMR in solids. Andrew et al.[38] and Lowe[39] first realized and

[38] E. R. Andrew, A. Bradbury, and R. G. Eades, *Arch. Sci.* **11**, 223 (1958).
[39] I. J. Lowe, *Phys. Rev. Lett.* **2**, 285 (1959).

demonstrated experimentally that it is possible to eliminate the secular parts of second-rank spin interactions in rigid solids by rapidly spinning the sample about an axis tilted by the "magic angle" $\beta_m = \arccos(1/\sqrt{3}) = 54°44'$ to the applied magnetic field. This possibility is probably highly astonishing for anyone with an unspoiled mind in view of the fact that the assertion is that rotation of the sample about one single axis is sufficient to suppress, e.g., all secular dipolar interactions of the spins, regardless of how the internuclear vectors $\mathbf{r}_{ik}$ are oriented with respect to the applied field.

One might suspect that this possibility rests on the fact that the dipolar interaction of a spin pair i, k is axially symmetric about $\mathbf{r}_{ik}$. However, this is not the case. It turns out that all (the secular parts of) second-rank tensor interactions are wiped out by magic-angle sample spinning, regardless of whether or not η is zero.

In order fully to appreciate magic-angle sample spinning let us consider the effect on the secular dipolar Hamiltonian of fast spinning the sample about an axis inclined by an arbitrary angle β to the applied field.

Let us start by considering $\mathscr{H}_D$ in its basic form

$$\mathscr{H}_D = -2\hbar \sum_{i<k} \gamma_n{}^i \gamma_n{}^k \sum_{m=-2}^{+2} (-1)^m R_{2,-m}^{D,\,ik} T_{2m}^{D,\,ik}. \tag{4-4}$$

This form of $\mathscr{H}_D$ may seem unnecessarily clumsy, but shortly it will be seen to be very efficient. Invoking, but not repeating, the arguments preceding Eq. (3-3) we write

$$R_{2,-m}^{D,\,ik}(\text{LABS}) = \sum_{m'} \mathscr{D}_{m',-m}^2(\Omega'') R_{2m'}^{D,\,ik}(\text{CRS})$$

$$= \sum_{m'} \mathscr{D}_{m',-m}^2(\Omega'') \sum_{m''} \mathscr{D}_{m''m'}^2(\Omega_{ik}') \rho_{2m''}^{D,\,ik}. \tag{4-5}$$

That is, we express the $R_{2,-m}^{D;\,ik}$ in the laboratory frame—where we need them—by

  (i)   the $\rho_{2m''}^{D,\,ik}$, which are rotationally invariant quantities,
  (ii) the Euler angles Ω_{ik}', which relate the arbitrarily chosen sample fixed frame (CRS) with the i, k-dipolar principal axes system, and
  (iii) the *one* set of Euler angles Ω'', which relate the laboratory and sample fixed frames.

Ω_{ik}' is obviously different for different spin pairs in the sample. Spinning the sample renders Ω'' time dependent: $\Omega'' = (0, \beta'', \omega_r t)$.

We are now ready to discuss the effects of sample spinning: Restriction to secular terms leaves from the sum over m only the term $m = 0$; hence the only $\mathscr{D}_{m'm}^2(\Omega'')$ that enter are the $\mathscr{D}_{m'0}^2(0, \beta'', \omega_r t) = \cdots e^{im'\omega_r t}$ (the dots mean "irrelevant at the moment"). It follows that restriction to time averages leaves us with $\mathscr{D}_{00}^2(0, \beta'', \ldots) = \frac{1}{2}(3\cos^2 \beta'' - 1)$. The remaining sum over m''

(which actually contains one term only because $\rho_{2m''}^{D,ik} = 0$ for $m'' \neq 0$) is thus multiplied by the factor $\frac{1}{2}(3\cos^2\beta'' - 1)$.

Let us consider two special cases:

(a) $\beta'' = \frac{1}{2}\pi$ (standard sample spinning about an axis perpendicular to the applied field). $\mathscr{D}_{00}^2(0, \frac{1}{2}\pi, \ldots) = -\frac{1}{2}$ and

$$\mathscr{H}_{\text{secular, effective}}^{D} = +\hbar \sum_{i<k} \gamma_n{}^i \gamma_n{}^k \mathscr{D}_{00}^2(\Omega'_{ik}) \sqrt{\tfrac{3}{2}}\, r_{ik}^{-3} T_{20}^{D,ik}$$

$$= +\tfrac{1}{2}\hbar \sum_{i<k} \gamma_n{}^i \gamma_n{}^k \tfrac{1}{2}(3\cos^2\beta'_{ik} - 1) r_{ik}^{-3}(3I_0{}^i I_0{}^k - \mathbf{I}^i\cdot\mathbf{I}^k).$$

(4-6)

For comparison purposes we also write

$$\mathscr{H}_{\text{secular}}^{D} = -\hbar \sum_{i<k} \gamma_n{}^i \gamma_n{}^k \tfrac{1}{2}(3\cos^2\beta_{ik} - 1) r_{ik}^{-3}(3I_0{}^i I_0{}^k - \mathbf{I}^i\cdot\mathbf{I}^k). \qquad (4\text{-}7)$$

The effective secular dipolar Hamiltonian that is operative under standard sample-spinning conditions thus differs from $\mathscr{H}_{\text{secular}}^{D}$ only in two subtle respects: First, β'_{ik}, which relates $\mathbf{r}_{ik}$ to the sample fixed frame (CRS), enters Eq. (4-6), whereas β_{ik}, which relates $\mathbf{r}_{ik}$ to the laboratory frame in the static case, enters Eq. (4-7). The difference is irrelevant for powder samples. Second, sample spinning enters an extra factor $-\frac{1}{2}$. The change of sign has no consequence upon the line broadening caused by $\mathscr{H}_{\text{secular, effective}}^{D}$; however a similar change of sign of an effective Hamiltonian, brought about in a somewhat different manner, has inspired the MIT group of J. Waugh to an interesting series of papers about time reversal experiments[40, 41] (see also Chapter VI, Section D).

(b) $\beta'' = \beta_m = \arccos(1/\sqrt{3}) = 54°44'08''$ (magic-angle sample spinning). The factor

$$\mathscr{D}_{00}^2(0, \beta_m, \omega_r t) = \tfrac{1}{2}(3\cos^2\beta_m - 1)$$

vanishes and so does $\mathscr{H}_{\text{secular, effective}}^{D}$. This is what Andrew et al.[38] and Lowe[39] first noticed and demonstrated. (Note that we arrived at this result with practically no calculations!)

Let us now inquire how magic-angle sample spinning affects the nuclear magnetic shielding, the indirect spin–spin coupling, and the quadrupolar Hamiltonians. It is clear that it does not affect the scalar constituents $(l = 0)$ of both $\boldsymbol{\sigma}$ and $\mathbf{J}$. Eventual constituents $(l = 1)$ are briefly mentioned below. The most important and the most interesting constituents are those with $l = 2$.

It is now very important to note that on the level of Eqs. (4-4) and (4-5)

[40] W.-K. Rhim, A. Pines, and J. S. Waugh, *Phys. Rev. Lett.* **25**, 218 (1970).
[41] W.-K. Rhim, A. Pines, and J. S. Waugh, *Phys. Rev. B* **3**, 684 (1971).

all the internal Hamiltonians $\mathscr{H}_D$, $\mathscr{H}_Q$, $\mathscr{H}_{CS}$ ($l = 2$), $\mathscr{H}_J$ ($l = 2$), have exactly the same form. They differ in the explicit forms of the T_{2m}^{λ}, $\lambda = D, Q, CS, J$, in the values of the parameters $\rho_{2m''}^{\lambda}$, and in the sets of the Euler angles Ω'. They agree with each other, in particular, in the occurrence of the sum $\sum_{m'} \mathscr{D}_{m',-m}^2 (\Omega'')$ in Eq. (4-5), with one and the same set of angles $\Omega'' = \alpha'', \beta'', \gamma''$ for all internal Hamiltonians.

The conclusions we have drawn for $\mathscr{H}_D$ were based on (1) restriction to secular terms ($m = 0$) and (2) restriction to time-independent terms ($m' = 0$). Both affect the sum $\sum_{m'} \mathscr{D}_{m',-m}^2 (\Omega'')$, which is common to all internal Hamiltonians. Hence, we may conclude, without any further calculations, that magic-angle sample spinning not only eliminates $\mathscr{H}_{secular}^D$ but also $\mathscr{H}_{secular}^{(2)CS}$, $\mathscr{H}_{secular}^{(2)J}$, and $\mathscr{H}_{secular}^Q$ provided restriction to secular terms is possible and provided the spinning speed is high enough. These two conditions have to be checked, of course, in each particular case and very often either one or both will not be fulfilled for $\mathscr{H}_Q$.[42]

By now it should have become clear that the magic angle β_m should actually be labeled with an index l. In fact there are "magic angles" for each rank l. It is obvious that $\mathscr{D}_{00}^l (0, \beta_m^{(l)}, ...) = P^l(\cos \beta_m^{(l)}) = 0$, where $P^l(\cos \beta)$ is the lth Legendre polynomial, defines the lth rank magic angle(s) $\beta_m^{(l)}$. For $l = 1$ we have, e.g., $\beta_m^{(1)} = \frac{1}{2}\pi$, which is, in fact, one of the magic angles for all cases $l =$ odd. It is the only one for $l = 1$. We thus conclude that sample spinning about the second-rank magic angle $\beta_m^{(2)}$ does not wipe out eventual antisymmetric constituents of σ and J.

The achievement of (second-rank) magic-angle sample spinning is thus the suppression of the dominant anisotropic ($l = 2$) interactions of spins in solids. As a result, isotropic interactions of the spins, in particular isotropic chemical shifts, do become measurable, or do so with much greater precision. It is not true in general that these parameters can also be measured in liquids and with much greater ease and accuracy: There are cases where the very event of crystallization creates chemical shifts or chemical-shift differences. A famous example is ZnP_2, which has been investigated in pioneering papers by Andrew and Wynn[43] and by Kessemeier and Norberg.[44]

Scalar indirect spin–spin couplings between spins on different molecules,

[42] Quadrupolar Hamiltonians can be analyzed *simply* in two limiting regimes only: $\|\mathscr{H}_Q\| \gg \|\mathscr{H}_Z\|$ (weak-field case) and $\|\mathscr{H}_Q\| \ll \|\mathscr{H}_Z\|$ (strong-field case). $\|\cdots\|$ means "size of ...." Our discussion obviously applies for the strong-field case only. It seems that rapid sample spinning has not been attempted or even discussed in the weak- (or zero-) field case. It is unclear what purpose sample spinning might serve in this regime. A discussion of such experiments would require additional considerations such as whether or not the spin magnetization can follow adiabatically the motion of the sample.

[43] E. R. Andrew and V. T. Wynn, *Proc. Roy. Soc., Ser. A* **291**, 257 (1966).

[44] H. Kessemeier and R. E. Norberg, *Phys. Rev.* **155**, 321 (1967).

unobservable in both isotropic and anisotropic liquids, do become measurable in principle. So far, however, experimental results are still on a rather speculative level.[44]

Likewise, antisymmetric constituents of **J** do become observable in principle. However, our pessimistic discussion of the measurability of anti-symmetric constituents of **σ** also applies to antisymmetric constituents of **J**, since none of the irreducible tensor components $T_{1m}^{J,ik}$ shifts any of the spin levels $|++\rangle$, $(1/\sqrt{2})(|-+\rangle+|+-\rangle)$, $|--\rangle$ $(I_i=\frac{1}{2}, I_k=\frac{1}{2}, I_{tot}=1)$ in first order. $T_{10}^{J,ik} = (1/\sqrt{2})(I_1^{i}I_{-1}^{k} - I_{-1}^{i}I_1^{k})$ *does* shift the $(1/\sqrt{2})(|-+\rangle-|+-\rangle)$ level $(I_i=\frac{1}{2}, I_k=\frac{1}{2}, I_{tot}=0)$ in first order. This level, however, is not involved in magnetic transitions. Therefore, it is no surprise that conclusive magic-angle sample-spinning experiments pertaining to antisymmetric constituents of indirect spin–spin couplings still seem to be lacking.

We mentioned above that the effects of magic-angle sample spinning are highly astonishing. Let us now look back and ask: What is the gist of the technique? It is that actually two rotations about two noncolinear axes are imposed simultaneously on the system. One is in ordinary space and leads to discarding all terms with $m' \neq 0$. The other is in spin space. We have referred to it as "restriction to secular terms," which meant discarding all terms with $m \neq 0$. In Section C we shall see that this is equivalent to a fast rotation about the z-axis in spin space, keeping only time-independent terms. For interactions of a given rank the only element of the $\mathscr{D}_{m'm}^{l}(\Omega'')$ matrix that survives is $\mathscr{D}_{00}^{l}(\Omega'')$. It is made to vanish by a clever choice of the angle between the two rotation axes. What counts is which terms are time independent and which are not. Relations between $\mathbf{B}_{st}$ and "unique axes" such as $\mathbf{r}_{ik}$ are completely irrelevant.

C. Averaging in Spin Space

We outline in this section two illustrative examples. In Section D we shall discuss the theory (called average Hamiltonian theory) in a more general way. Applications of the average Hamiltonian theory to high-resolution NMR in solids and fluids conclude this chapter (Sections E and F). The two intro-ductory examples are the famous Van Vleck truncation of internal Hamiltonians (1948),[45] and the WAHUHA four-pulse experiment (WAugh, HUber, and HAeberlen, 1968).[46]

1. TRUNCATION OF INTERNAL HAMILTONIANS

As stated in the introduction of the previous section an appropriate inter-action representation that renders the operators T_{lm}^{λ} time dependent is a

[45] J. H. Van Vleck, *Phys. Rev.* **74**, 1168 (1948).
[46] J. S. Waugh, L. M. Huber, and U. Haeberlen, *Phys. Rev. Lett.* **20**, 180 (1968).

prerequisite for averaging in spin space. We said that interaction representations that put fast common motions of the spins (or groups thereof) on the operators qualify as appropriate.

The Zeeman interaction of the spins with a large applied magnetic field leads to a fast precession of the spins. The speed of this precession is common to all spins of a given isotope. An interaction representation that removes the Zeeman term from the total spin Hamiltonian thus qualifies as appropriate in our sense. This interaction representation is introduced by the following steps:

The total spin Hamiltonian is

$$\mathscr{H} = \mathscr{H}_Z + \mathscr{H}^{\text{int}}. \tag{4-8}$$

The equation of motion of the spin density matrix ρ is

$$\dot{\rho}(t) = -i[\mathscr{H}, \rho(t)]. \tag{4-9}$$

Note that $\mathscr{H}$ is time independent as far as the operators are concerned.

We now make the *ansatz*

$$\rho(t) = U_Z \rho_R(t) U_Z^{-1} \quad \text{with} \quad U_Z = \exp[-i\mathscr{H}_Z t], \tag{4-10}$$

which leads to

$$
\begin{aligned}
\dot{\rho} &= \dot{U}_Z \rho_R U_Z^{-1} + U_Z \dot{\rho}_R U_Z^{-1} + U_Z \rho_R \dot{U}_Z^{-1} \\
&= -i\mathscr{H}_Z U_Z \rho_R U_Z^{-1} + U_Z \dot{\rho}_R U_Z^{-1} + iU_Z \rho U_Z^{-1} \mathscr{H}_Z \\
&= \underbrace{-i[\mathscr{H}_Z, U_Z \rho_R U_Z^{-1}] + U_Z \dot{\rho}_R U_Z^{-1}}_{\text{left side of Eq. (4-9)}} = \underbrace{-i[(\mathscr{H}_Z + \mathscr{H}^{\text{int}}), U_Z \rho_R U_Z^{-1}]}_{\text{right side of Eq. (4-9)}}.
\end{aligned}
$$

The commutators with $\mathscr{H}_Z$ cancel and we are left with

$$U_Z \dot{\rho}_R U_Z^{-1} = -i[\mathscr{H}^{\text{int}}, U_Z \rho_R U_Z^{-1}].$$

Multiplying from the left with U_Z^{-1} and from the right with U_Z gives

$$\dot{\rho}_R = -i\{\underbrace{U_Z^{-1} \mathscr{H}^{\text{int}} U_Z}_{\mathscr{H}_R^{\text{int}}} \rho_R \underbrace{U_Z^{-1} U_Z}_{1} - \underbrace{U_Z^{-1} U_Z}_{1} \rho_R \underbrace{U_Z^{-1} \mathscr{H}^{\text{int}} U_Z}_{\mathscr{H}_R^{\text{int}}}\}$$

$$= -i[\mathscr{H}_R^{\text{int}}, \rho_R], \tag{4-11}$$

which is the equation of motion of the density matrix (von Neumann equation) in the interaction representation defined by Eq. (4-10). The relevant Hamiltonian is

$$\mathscr{H}_R^{\text{int}}(t) = U_Z^{-1}(t)\mathscr{H}^{\text{int}} U_Z(t). \tag{4-12}$$

The factors of $\mathscr{H}^{\text{int}}$ on which U_Z and U_Z^{-1} operate are the spin factors contained in the T_{lm}^λ:

$$T_{lm}^\lambda \to T_{R,lm}^\lambda = U_Z^{-1} T_{lm}^\lambda U_Z.$$

We have now reached a point where it seems appropriate to make a number of comments.

a. *Truncation*

$\mathcal{H}^{int}$ becomes time dependent when sandwiched between U_Z^{-1} and U_Z. Some of the T_{lm}^λ acquire a periodic time dependence, others do so in part, and still others remain completely time independent. Truncation of $\mathcal{H}_R^{int}$ means rejection of all those terms whose operators depend on time in such a way that their time averages vanish. We have, in fact, restricted ourselves repeatedly to truncated internal Hamiltonians in the preceding sections.

In order to find the time-independent operators of $\mathcal{H}_R^{int}$ we recall the following well-known relations (see, e.g., Abragam,[9] p. 23):

$$e^{i\omega t I_0} I_{\pm 1} e^{-i\omega t I_0} = I_{\pm 1} e^{\pm i\omega t}, \tag{4-13a}$$

$$e^{i\omega t I_0} I_0 e^{-i\omega t I_0} = I_0. \tag{4-13b}$$

(Recall $I_0 \equiv I_z$.) Applying them to the T_{lm}^λ operators of Table 2-2 results in a periodic time dependence of all $T_{R,lm}^\lambda$ with $m \neq 0$; (exception: $T_{R,2\pm1}^{SR}$; these operators, however, will be of no further relevance in our context). The $T_{R,l0}^\lambda$ remain time independent. The $T_{R,l0}^{D;ik}$ and $T_{R,l0}^{J;ik}$ remain completely time independent only when i and k represent like spins (which means spins with identical magnetogyric ratios). When spins i and k are unlike, only the parts $\left(1/\sqrt{6}\right)\left(2I_0{}^i I_0{}^k\right)$ out of $T_{20}^{D,J;ik} = \left(1/\sqrt{6}\right)\left(3I_0{}^i I_0{}^k - \mathbf{I}^i\cdot\mathbf{I}^k\right)$ and $I_0{}^i I_0{}^k$ out of $T_{00}^{J,ik} = \mathbf{I}^i\cdot\mathbf{I}^k$ remain time independent. This completes the list of terms that are *not* rejected by truncation.

b. *Equivalence of Truncation and First-Order Perturbation Theory*

It is obvious from the generation of the $T_{R,lm}^\lambda$ that truncation retains only such terms of $\mathcal{H}_R^{int}$ whose operators commute with $\mathcal{H}_Z$. When we treat $\mathcal{H}^{int}$ as a perturbation of $\mathcal{H}_Z$ it is exactly the same set of terms that causes first-order energy level shifts, and hence first-order spectral line shifts and splittings. Truncation and first-order perturbation theory are equivalent.

c. *Rotating Spin Frame*

Consider for the moment a spin system consisting of like spins only. $I_z = \sum_i I_z{}^i$ is the operator of infinitesimal rotations about the z-axis in spin space. U_Z, therefore, has the effect of a transformation into a rotating spin frame. This rotating spin frame is accelerated with respect to the laboratory frame. Inertial forces (or moments) appear and just cancel the physical moments exerted on the spins by $\mathbf{B}_{st}$.

Forces (expressed by Hamiltonians) that are static in the laboratory frame appear explicitly time dependent when viewed from the rotating spin frame: $\mathcal{H}^{int} \rightarrow \mathcal{H}_R^{int}(t)$. Time dependences introduced in this way into the relevant

Hamiltonian cannot and should not be distinguished from time dependences arising from physical motions of the molecules. We can treat them alike, that is, we can reject terms with zero time averages in lowest order analyses of NMR spectra. This procedure obviously covers truncation but is, in fact, more general.

Viewing matters from a rotating spin frame is a geometric interpretation of the interaction representation defined by $U_Z = \exp[-i\mathcal{H}_Z t]$. This interpretation leads us to introduce *expectation values in the rotating frame* of observables O, defined by

$$\langle O \rangle_R = \text{tr}\, \rho_R\, O. \tag{4-14}$$

Note that O and *not* O_R enters the r.h.s. of Eq. (4-14)! The expectation value of the observable O in the laboratory frame is given by

$$\text{tr}\, \rho_R\, O_R = \text{tr}\{U_Z^{-1}\rho U_Z\, U_Z^{-1} O U_Z\} = \text{tr}\, \rho O.$$

As an example suppose $\rho(0) = \rho_R(0) \propto (1 + I_x)$. Such an initial density matrix can be prepared from a thermal equilibrium situation by a so-called 90° pulse. Suppose, further, that $\mathcal{H}^{\text{int}}$ is so small that it can be neglected during the observation time of the system. The expectation value of $\mathbf{M}$ in the laboratory frame is rotating with angular velocity $\omega_0 = \gamma B_{\text{st}}$ about $\mathbf{B}_{\text{st}}$; the expectation value of $\mathbf{M}$ in the rotating frame, however, is stationary—$\mathbf{M}$ is seen stationary by an observer placed into the rotating spin frame.

When NMR people try to imagine what is happening in a spin system that is subjected to, say, a certain sequence of pulses, they actually think in terms of expectation values in the rotating frame. The NMR signal after phase-sensitive detection—preferably quadrature phase-sensitive detection[47,48]—is directly proportional to the expectation values in the rotating frame of $I_x = \sum_i I_x^i$ and $I_y = \sum_i I_y^i$.

If the spin system consists of two species of nuclear spins with, say, gyromagnetic ratios γ_n^I and γ_n^S,

$$U_Z = \exp[-i\mathcal{H}_Z t] = \exp[it(\omega_0^I I_z + \omega_0^S S_z)]$$
$$= \exp[i\omega_0^I I_z t] \exp[i\omega_0^S S_z t]$$

can be interpreted as effecting a transformation into two rotating spin frames, one for the I spins, the other for the S spins, rotating with different angular velocities ω_0^I and ω_0^S about the applied field. Having accepted that space and spin variables are described in different frames it should cause no extra difficulty that spin variables of different species are described in different frames.

[47] J. D. Ellett, Jr., M. G. Gibby, U. Haeberlen, L. M. Huber, M. Mehring, A. Pines, and J. S. Waugh, *Advan. Magn. Resonance* **5**, 117 (1971).
[48] A. G. Redfield and R. K. Gupta, *Advan. Magn. Resonance* **5**, 82 (1971).

d. *Rotations in Spin Space—Behavior of T_{lm} Operators*

With magnetic fields, both static and oscillatory, we can impose rotations upon the spins. By treating these fields by interaction representations we transfer the corresponding rotations upon the T_{lm} operators. Naturally, only the spin parts of these operators are affected. If we apply rf fields at or close to resonance with the Larmor frequency of one particular spin species, say the I-spin species, only this particular spin species is affected. This all means that the T_{lm}^{λ} operators do not transform under rotations in the subspace of the I spins as irreducible tensor operators of rank l and row m.

Table 4-1 shows how *those* T_{lm}^{λ} operators that are not rejected by truncation transform in the I-spin subspace. It is of crucial importance for the success of selective averaging in spin space that the ranks l of T_{00}^{CS}, T_{20}^{CS}, $T_{20}^{D}(I, S)$, $T_{00}^{J}(I, S)$, $T_{20}^{J}(I, S)$ be different in full space and in I-spin subspace. It is this difference that provides the key for, e.g., discriminating between the truncated shielding and the truncated like-spin dipolar Hamiltonians. While it is possible to discriminate between these Hamiltonians, it is impossible to discriminate between the direct (dipolar) and the indirect spin–spin coupling Hamiltonians: they behave identically under rotations in both ordinary and spin spaces and, as a result, they "live" and "die" together.

However, this statement should not be unduly overgeneralized. It applies for discrimination by *motions* in both ordinary and spin spaces. (By "discrimination" we mean the performance of an experiment in which one type of interaction has a *zero* effect, whereas another has a *nonzero* effect.) The fact that the i, k dipole interaction is axially symmetric about the internuclear vector $\mathbf{r}_{ik}$, whereas the indirect i, k spin–spin interaction is not—at least not

TABLE 4-1

RANK l AND ROW m ACCORDING TO WHICH THE OPERATORS OF $\mathscr{H}^{int}$ LEFT OVER BY THE
TRUNCATION PROCESS TRANSFORM UNDER ROTATIONS IN I-SPIN SUBSPACE

Interaction	T_{l0}^{λ}	After truncation	Transforms in I-spin space according to
Shielding	T_{00}^{CS}	$I_0{}^i B_0$	$l = 1, m = 0$
	T_{20}^{CS}	$\sqrt{\tfrac{2}{3}}\, I_0{}^i B_0$	$l = 1, m = 0$
Quadrupolar	T_{20}^{Q}	$(1/\sqrt{6})[3(I_0{}^i)^2 - (\mathbf{I}^i)^2]$	$l = 2, m = 0$
Direct and indirect between $\mathbf{I}^i$ and $\mathbf{I}^k$	T_{00}^{J}	$\mathbf{I}^i \cdot \mathbf{I}^k$	$l = 0, m = 0$
	$T_{20}^{D,J}$	$(1/\sqrt{6})[3 I_0{}^i I_0{}^k - \mathbf{I}^i \cdot \mathbf{I}^k]$	$l = 2, m = 0$
Direct and indirect between $\mathbf{I}^i$ and $\mathbf{S}^k$	T_{00}^{J}	$I_0{}^i S_0{}^k$	$l = 1, m = 0$
	$T_{20}^{D,J}$	$\sqrt{\tfrac{2}{3}}\, I_0{}^i S_0{}^k$	$l = 1, m = 0$

necessarily—could be exploited in the following type of experiment: One orients a single crystal, which contains pairs of magnetically well-isolated, magnetically equivalent spins $I = \frac{1}{2}$, such that the internuclear vectors r_{ik} are at an angle of $\beta_m = 54°44''$ to the applied field. If there is no indirect spin–spin coupling no splitting of the NMR line of this pair of nuclei will be observed. If there is indirect spin–spin coupling with $\mathbf{J}$ not axially symmetric about r_{ik} a splitting of the NMR line will be observed. One problem of such an experiment is obviously the orientation of the crystal. One must rely on structural information from other sources than NMR from that pair of nuclei. To our knowledge no such experiment has been reported so far.

After this digression let us return to averaging in spin space and let us discuss the WAHUHA four-pulse experiment. This experiment discriminates between truncated shielding and truncated like-spin dipolar interactions. It will serve us to discuss further subtle features of averaging in spin space.

2. THE WAHUHA FOUR-PULSE EXPERIMENT

What we need to bring into play now are rf fields. Without mentioning it we have needed them so far to observe NMR spectra, either by kicking spin systems off their thermal equilibrium states—the observation of the return of the systems into thermal equilibrium could then be done in the absence of applied rf fields—or by cautiously "tickling" spin systems with such small rf fields that they are not perturbed significantly.

In what follows rf fields have the additional task of modifying NMR spectra. It is important to realize this double role[49] of the rf fields in multiple-pulse and spin-decoupling experiments.

a. *rf and Resonance Offset Hamiltonians; Quadrature Phase rf Pulses*

In a WAHUHA four-pulse experiment the spin system is irradiated with a sequence of rf pulses, the amplitudes, widths, spacings, and carrier phases of which are monitored. We postpone a more detailed description of the sequence until we are ready to give it in the rotating frame. Let us suppose for the moment that the spin system consists of like spins only. The rf Hamiltonian is then

$$\mathcal{H}_{rf} = -\gamma_n{}^I B_1(t) \cos(\omega t + \varphi(t)) \sum_i I_x{}^i$$

$$= -\tfrac{1}{2}\gamma_n{}^I B_1(t) \{\exp[i(\omega t + \varphi(t))]$$

$$+ \exp[-i(\omega t + \varphi(t))]\}(1/\sqrt{2})(I_{-1} - I_{+1}).$$

---

[49] In some of the dynamic polarization experiments[50] rf fields even have a triple role: (1) excitation, observation, (2) modifying spectra, and (3) establishing a thermal contact between different spin energy reservoirs.

[50] A. Pines, M. G. Gibby, and J. S. Waugh, *J. Chem. Phys.* **56**, 1776 (1972).

We replaced $\sum_i I_x{}^i = I_x$ by $(1/\sqrt{2})(I_{-1} - I_{+1})$ [see Eqs. (2-21) and (2-22)].

Let us now partition the applied static field B_0 into $B_0{}^I + \Delta B$, where $B_0{}^I$ is chosen such that $-\gamma_n{}^I B_0{}^I = \omega$. We assume $\Delta \omega = |\gamma_n{}^I \Delta B| \ll |\omega|$; this has to be taken care of by the experimenter.

$\mathscr{H}_{\mathrm{off}} = -\Delta \omega I_z$, called the resonance offset Hamiltonian, has to be included in $\mathscr{H}^{\mathrm{int}}$. The rf Hamiltonian in the rotating frame defined by

$$U_Z = \exp\{+i\gamma_n{}^I B_0{}^I I_z t\} = \exp\{-i\omega I_z t\}$$

is

$$\mathscr{H}_R^{\mathrm{rf}} = -\tfrac{1}{2}\gamma_n{}^I B_1(t)\{e^{i[\omega t + \varphi(t)]} + e^{-i[\omega t + \varphi(t)]}\}(1/\sqrt{2})(I_{-1}e^{-i\omega t} - I_{+1}e^{i\omega t}).$$

In accordance with our general policy we reject high-frequency terms and are left with

$$\mathscr{H}_{R,\,\mathrm{eff}}^{\mathrm{rf}} = -\tfrac{1}{2}\gamma_n{}^I B_1(t)[I_x \cos \varphi(t) + I_y \sin \varphi(t)]. \tag{4-15}$$

We stress that the angular velocity of the rotating frame is chosen according to ω, which is a or the spectrometer frequency, and not according to a or the Larmor frequency of some spins in B_0.

In the WAHUHA experiment, and in many others, $\mathscr{H}_{\mathrm{rf}}$ consists of a series of pulses. $B_1(t)$ is rapidly switched on and off:

We abbreviate $\tfrac{1}{2}\gamma_n{}^I B_1$ by ω_1. A "90° pulse" is defined by $\omega_1 t_w = \tfrac{1}{2}\pi$.

Under resonance conditions ($\mathscr{H}_{\mathrm{off}} = 0$) it rotates ("flips") all I-spins through 90° in the rotating frame. A 90° δ pulse is a 90° pulse with $t_w \to 0$. It is an idealization since it requires $B_1 \to \infty$.

90° pulses, either real or δ pulses, are called x, $-x$, y, $-y$ pulses, when during the pulse $\varphi(t)$ is 0, π, $\tfrac{1}{2}\pi$, $\tfrac{3}{2}\pi$, respectively. x, $-x$, y, and $-y$ pulses are quadrature-phase rf pulses. We shall refer to them as P_x, P_{-x}, P_y, and P_{-y} pulses. Under resonance conditions they rotate the I-magnetization in the rotating I-spin frame about the $+x$, $-x$, $+y$, and $-y$ axes, respectively, through 90°. We do not need to elaborate here whether the rotations are clockwise or anticlockwise.

b. WAHUHA Sequence with δ pulses; Average Hamiltonian Approximation

The WAHUHA experiment is a special Fourier transform (FT) NMR experiment. In a standard FT experiment [Fig. 4-1(1)] the spin system is excited with a short intense rf pulse, say a P_x pulse. It then evolves according to $\mathscr{H}_{\mathrm{secular}}^{\mathrm{int}}$. While it does so it induces a voltage, called *free induction decay* (FID) in the receiver coil of the spectrometer. The FID is phase-sensitively

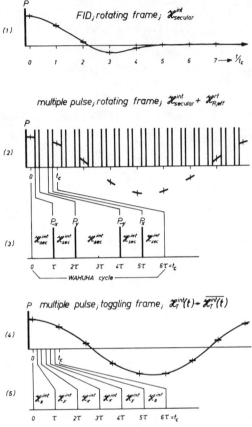

FIG. 4-1. Principles of resolution enhancement by multiple-pulse sequences; analogy to Fourier transform NMR.

(1) Free induction decay (FID) of nuclear magnetization $\mathbf{M}$ after preparation pulse P. Samples (indicated by short heavy lines) of the signal voltage are taken at regular intervals t_c and Fourier transformed. The resulting spectrum corresponds to $\mathscr{H}_{\text{secular}}^{\text{int}}$.

(2) Multiple-pulse experiment. Extra pulses are applied between each pair of "samples." The gaps between the pulses are called "windows." The signal voltage jumps wildly from one window to the next. Windows where "samples" of the signal are taken are called "observation windows." All events are described in the rotating spin frame. The spin system evolves under the influence of $\mathscr{H}_{\text{secular}}^{\text{int}} + \mathscr{H}_{\text{R, eff}}^{\text{rf}}$.

(3) Particular pulse sequence of the WAHUHA experiment. The same sequence is repeated over and over. The period is t_c.

(4) Same events viewed from the toggling frame. Except for P no more pulses appear. However,

(5) $\mathscr{H}_{\text{secular}}^{\text{int}}$ has turned into a periodically time-dependent form: $\mathscr{H}_{\text{T}}^{\text{int}}(t)$. Provided t_c is small enough $\mathscr{H}_{\text{T}}^{\text{int}}(t)$ can be replaced by its time average, $\overline{\mathscr{H}_{\text{T}}^{\text{int}}(t)}$. In particular, $\overline{\mathscr{H}_{\text{T}}^{\text{D}}(t)} = 0$, $\overline{\mathscr{H}_{\text{T}}^{\text{CS}}(t)} \neq 0$.

Rotating and toggling frames coincide during the observation windows. The smooth path of $\mathbf{M}$ in the toggling frame can be observed stroboscopically in the rotating frame. The spectrum finally obtained corresponds to $\overline{\mathscr{H}_{\text{T}}^{\text{int}}(t)}$.

detected. The frequency of the local oscillator of the phase-sensitive detector (PSD) is ω. The output of the PSD is an audio signal, the frequencies of which depend on $\mathcal{H}^{int}_{secular}$ including $\mathcal{H}_{off}$. "Samples" of the phase-sensitively detected FID are taken at regular time intervals t_c. They are digitized and subsequently fourier transformed. The result is the spectrum of the spin system, which corresponds to $\mathcal{H}^{int}_{secular}$.

A line-narrowing multiple-pulse experiment works in exactly the same way as far as excitation and observation of the spin system, and data handling are concerned. However, between the "sample" points the spin system is irradiated with extra pulses [Fig. 4-1 (2)]. The particular sequence of pulses employed in the WAHUHA sequence is shown in Fig. 4-1 (3).

The spin system evolves now according to $\mathcal{H}^{int}_{secular} + \mathcal{H}^{rf}_{R, eff}$. $\mathcal{H}^{rf}_{R, eff}$ imposes on the I spins a common motion (as did $\mathcal{H}_Z$), which we again treat by way of an interaction representation ansatz $\rho_R(t) = U_{rf} \rho_T U_{rf}^{-1}$, where T stands for "toggling" frame. This notation will become clear shortly. What is new with U_{rf} in comparison with U_Z is the time dependence of the relevant Hamiltonian, $\mathcal{H}^{rf}_{R, eff}$. This Hamiltonian, however, is still constant in a finite number of time intervals (see Fig. 4-2). Therefore, we can go through a procedure completely

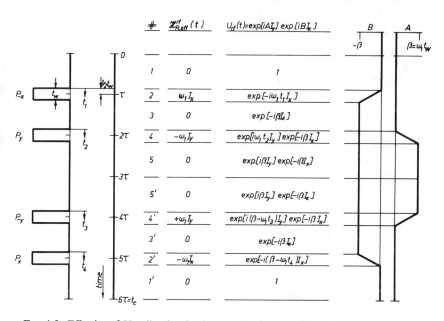

FIG. 4-2. Effective rf Hamiltonian in the rotating frame, $\mathcal{H}^{rf}_{R, eff}(t)$, for the WAHUHA sequence, and $U_{rf}(t)$. We distinguish ten intervals, 1–5 and 5′–1′. The graphs for A and B show that $U_{rf}(t)$ is symmetric about the midpoint 3τ of the cycle. $\beta = \omega_1 t_w$ is called "flip angle." For $t_w/\tau \to 0$ the optimum flip angle is $\beta_{opt} = \frac{1}{2}\pi$.

analogous to the one that led us to U_Z for each of these time intervals successively. The result can be written immediately and is also indicated in Fig. 4-2. Note that at the end of the cycle U_{rf} has returned to unity: U_{rf} is periodic modulo t_c. We shall return to this important property of multiple-pulse sequences in more detail in Section D.

Concentrating our attention on the "windows" of the pulse train we may now interpret U_{rf} as effecting a transformation into a "toggling frame." In the absence of $\mathscr{H}^{int}$ the spin density matrix ρ_T remains stationary in a frame that toggles with respect to the rotating frame in a rhythm imposed by the pulses of the sequence. Initially the toggling and rotating frames coincide. The first pulse tips the z axis of the toggling frame down to the y axis of the rotating frame. The x axes of the toggling and rotating frames remain parallel. The second pulse pushes the toggling frame z axis parallel to the rotating frame x axis, the third kicks it back to the rotating frame y axis, and the fourth restores the initial orientation of the toggling frame: $U_{rf}(t_c) = 1$. Then the game starts anew, and at all times $t = Nt_c$, N integer, we have $U_{rf} = 1$.

This means that the rotating and toggling spin frames coincide at $t = Nt_c$. They coincide, in fact, during the entire windows of the pulse train, which occur around Nt_c. It is in these windows ("observation windows") that we observe the nuclear magnetization.

We pointed out that with the aid of phase-sensitive detectors we can observe the nuclear spin magnetization **M** directly in the rotating frame. We cannot observe it so simply in the toggling frame. However, by observing **M** in the rotating frame, but restricting the observation to the "observation windows," we can actually observe **M** stroboscopically in the toggling frame. The electronic device used for that purpose is a "sample and hold" or "integrate and hold" unit.

Observing a nuclear magnetization stroboscopically is absolutely standard in FT NMR. What is peculiar of multiple-pulse experiments is that we are no longer free to choose the rhythm of opening and closing the gate of the sample and hold device: We are only allowed to open it during the proper windows of the sequence.

Let us now return to the density matrix in the toggling frame, ρ_T, which evolves according to

$$\dot{\rho}_T = -i[\mathscr{H}_T^{int}(t), \rho_T(t)], \tag{4-16}$$

with

$$\mathscr{H}_T^{int}(t) = U_{rf}^{-1}(t)\mathscr{H}_{secular}^{int} U_{rf}(t). \tag{4-17}$$

Since U_{rf} is periodic modulo t_c, $\mathscr{H}_T^{int}(t)$ is also periodic modulo t_c. Equations (4-16) and (4-17) say that the density matrix (the "spin system") evolves in the toggling frame according to a periodically time-dependent internal Hamiltonian $\mathscr{H}_T^{int}(t)$.

Alternatively, we may say that if we restrict the observation of the spin system in the rotating frame to the "observation windows," it behaves as if it were evolving in the absence of rf pulses according to $\mathscr{H}_T^{int}(t)$. This is shown in Fig. 4-1 (4, 5).

The final step is to take the time average of $\mathscr{H}_T^{int}(t)$ over its period t_c and to claim that the dominant term that governs the time evolution of the spin system in the toggling frame (in the rotating frame if $\mathbf{M}$ is observed stroboscopically) is the time-independent average Hamiltonian $\overline{\mathscr{H}_T^{int}(t)}$. The condition for this to hold is $\| t_c \mathscr{H}_{secular}^{int} \| \ll 1$.

This completes the analogy with a standard FT experiment. We summarize it as follows:

Standard FT: The spin system evolves in the rotating frame according to $\mathscr{H}_{secular}^{int}$, which is time independent. The spectrum calculated from a set of regularly spaced "sample points" of the FID corresponds to $\mathscr{H}_{secular}^{int}$.

WAHUHA multiple pulse: The spin system evolves in the toggling frame, and also in the rotating frame when the observation is <u>restricted</u> to the proper set of regularly spaced "sample points" according to $\overline{\mathscr{H}_T^{int}(t)}$, which again is time independent. The spectrum calculated by the computer corresponds to $\overline{\mathscr{H}_T^{int}(t)}$.

It remains for us to inspect $\overline{\mathscr{H}_T^{int}(t)}$. To keep matters simple we shall assume that all the pulses of the WAHUHA train are δ pulses. For $\mathscr{H}_T^{int}(t)$ we can then restrict ourselves to the contributions of $\mathscr{H}_T^{int}(t)$ from the *windows* of the pulse train. In Chapter V, Section D, we shall drop this restriction. The operators we need to consider are listed in Table 4-1. They can be classified according to their rank l in I-spin subspace.

$l = 0$ There is only one case: $\mathbf{I}^i \cdot \mathbf{I}^k$. This scalar product is left invariant by the rotations represented by $U_{rf}(t)$.

$l = 1$ There are four entries $l = 1$ in Table 4-1. All, however, boil down to $I_0 = I_z$ in the subspace of the I spins. Table 4-2 shows the transformations of I_z in the toggling frame.

$l = 2$ There are two entries $l = 2$ in Table 4-1. The more general one is $3I_z^i I_z^k - \mathbf{I}^i \cdot \mathbf{I}^k$. Table 4-2 also shows the transformations of this operator in the toggling frame.

The decisive point is that the average of spin operators with rank 2 vanishes, whereas the average of spin operators with rank 1 does not.

The WAHUHA sequence discriminates therefore between, in particular, like-spin dipolar interactions ($l = 2$) and isotropic as well as anisotropic chemical shielding ($l = 1$).

This is what Waugh first realized (1968) and this is what we demonstrated

TABLE 4-2

Transformations of I_z and $(3I_z^i I_z^k - \mathbf{I}^i \cdot \mathbf{I}^k)$ in the Toggling Frame during the Windows of the WAHUHA Sequence, and Averages Thereof for $t_W/\tau \to 0^a$

Time	U_{rf}	$U_{rf}^{-1} I_z U_{rf}$	$U_{rf}^{-1}(3I_z^i I_z^k - \mathbf{I}^i \cdot \mathbf{I}^k) U_{rf}$	$\mathcal{H}_T^{int}(t)$
$0 \cdots \tau,\, 5\tau \cdots 6\tau$	1	I_z	$3I_z^i I_z^k - \mathbf{I}^i \cdot \mathbf{I}^k$	$\mathcal{H}_z^{int}$
$\tau \cdots 2\tau,\, 4\tau \cdots 5\tau$	$\exp(-\tfrac{1}{2}i\pi I_x)$	I_y	$3I_y^i I_y^k - \mathbf{I}^i \cdot \mathbf{I}^k$	$\mathcal{H}_y^{in\flat}$
$2\tau \cdots 3\tau,\, 3\tau \cdots 4\tau$	$\exp(\tfrac{1}{2}i\pi I_y)\exp(-\tfrac{1}{2}i\pi I_x)$	I_x	$3I_x^i I_x^k - \mathbf{I}^i \cdot \mathbf{I}^k$	$\mathcal{H}_x^{int}$
average		$\tfrac{1}{3}(I_x + I_y + I_z)$	0	

^{a} The various forms of $\mathcal{H}_T^{int}(t)$ can be labeled conveniently by x, y, and z (last column).

in a series of papers about coherent averaging in spin space,[46, 51] which started the development of high-resolution NMR in solids by multiple-pulse techniques.

c. Comments

(i) *Variant choices of observation windows.* We have stressed repeatedly the importance of restricting the observation of the spin system to the "proper" windows of the pulse train. The windows around $t = 0, 6\tau, \ldots, N6\tau$ have been claimed to be the proper ones. However, there is little magic about this particular set of windows; all it does is the following:

(1) It makes the analogy between the WAHUHA and the FID experiments as perfect as possible and thus serves aesthetics.

(2) It makes $U_{rf}(t)$ and, as a result, $\mathcal{H}_T^{int}(t)$ symmetric about the center of the cycle.

This has no consequence on the level of only averages of Hamiltonians but will prove profitable when corrections are taken into account (Section D and Chapter V).

(ii) *How necessary are $U_{rf}(t)$ and the toggling frame?* It is clear that $U_{rf}(t)$ and the toggling frame are only theoretical aids for describing the outcome of experiments and are not essential to the experiments themselves. There is a tendency among some in the field to feel that their introduction is basically nothing but fuss, which can easily be circumvented by considering the motion of the expectation value $\langle \mathbf{I} \rangle_R$ of the nuclear spin angular momentum $\mathbf{I}$ in the rotating frame during the pulse train. In formulating the "argument" we shall use the typical jargon.

[51] U. Haeberlen and J. S. Waugh, *Phys. Rev.* 175, 453 (1968).

Initially the angular momentum is I_z. The first pulse flips it in the rotating frame to I_y, the second back to I_z, the third sends it to I_x (signs are disregarded), etc. On the average, the "argument" continues, we have

$$\langle \mathscr{H}_{CS} \rangle_{av} \propto \langle I_z \rangle_{av} = \tfrac{1}{3}(I_x + I_y + I_z) \quad \text{and} \quad \langle \mathscr{H}_{D} \rangle_{av} \propto \langle 3I_z{}^i I_z{}^k - \mathbf{I}^i \cdot \mathbf{I}^k \rangle_{av} = 0.$$

That is all.[52] The basic mistake is the confusion of spin operators, density matrix, and expectation values. While $\langle \mathbf{I} \rangle_R$ does vary with time in the rotating frame, the operator I_z does not. The rotating frame operator I_z varies only when viewed from the toggling frame.

How fallacious the "argument" just given is can be deomstrated most clearly by considering the following pulse train,[53] which in shorthand notation is represented by

$$P_x - \tau - \underbrace{\tau - P_{-y} - \tau - P_x - \tau - \tau - P_{-x} - \tau - P_y - \tau - \tau - P_{-y}}_{\text{repeated sequence}} - \cdots.$$

All pulses are 90° pulses. One glance at this pulse train will convince the reader that $\langle \mathbf{I} \rangle_R$ always points along either the $-y$ or the $-z$ axis of the rotating frame, but never along the x axis. The reasoning outlined above does *not*, therefore, lead, e.g., to $\langle \mathscr{H}_D \rangle_{av} = 0$. Yet this pulse train is perfectly equivalent to the WAHUHA pulse train as far as averaging of nuclear shielding and like-spin dipolar Hamiltonians is concerned. It simply has two extra preparation pulses, which cannot and do not affect operator averages.

d. NMR Signal during WAHUHA Sequence; Scaling Factor

Let us consider the NMR signal that appears as a voltage at the output of the phase-sensitive detector (PSD) (detectors if quadrature PSDs are used) before it is sent through the sample and hold unit(s). This topic is somewhat off the track we have been following so far; it is, however, very important for anyone who faces the delicate task of aligning a WAHUHA or another line-narrowing multiple-pulse sequence.

The first steps of the alignment procedure are usually done on a liquid sample whose NMR spectrum consists of a single line only (e.g., water or hexafluorobenzene). The truncated shielding and the offset Hamiltonians have the same spin operator form. The offset Hamiltonian is under control of the experimenter and can therefore be used to simulate different chemical shifts. Extensive use is made of this possibility in every multiple-pulse laboratory. In what follows let us assume that the genuine chemical shift—there is only one—of the test sample has been absorbed in the definition of $\gamma_n{}^I B_0{}^I$ so that

---

[52] The author admits having used this false "argument" himself when trying to explain in a hurry the principles of high-resolution NMR in solids to visitors in his laboratory. He wishes to apologize on this occasion.

[53] P. Mansfield, *Phys. Lett. A* **32**, 485 (1970).

the terms "above," "on," and "below" resonance refer only to the offset Hamiltonian ($\Delta B > 0$, $=0$, <0). It will also be convenient to define x- and y-PSD's. The phase of the reference signal of an x-PSD is chosen such that the output of the PSD is proportional to the component of the nuclear magnetization along the x axis of the rotating frame:

$$\text{output of } x\text{-PSD} \propto \langle M_x \rangle_R \propto \text{tr}\, \rho_R I_x,$$

and similarly

$$\text{output of } y\text{-PSD} \propto \langle M_y \rangle_R \propto \text{tr}\, \rho_R I_y.$$

During a WAHUHA pulse train in the absence of $\mathscr{H}^{\text{int}}$ including $\mathscr{H}_{\text{off}}$ the density matrix $\rho_R(t)$ evolves according to

$$\rho_R(t) = U_{\text{rf}}(t)\, \rho_T(t)\, U_{\text{rf}}^{-1}(t),$$

where

$$\rho_T(t) = \text{const} = \rho_R(0) \propto \exp(\tfrac{1}{2}i\pi I_x)\, I_z \exp(-\tfrac{1}{2}i\pi I_x) = I_y.$$

(Those parts of ρ_R proportional to $\mathbb{1}$ in spin space are irrelevant for the NMR signal; we ignore them.)

$U_{\text{rf}}(t)$ is given in Fig. 4-2. The absence of $\mathscr{H}^{\text{int}}$ including $\mathscr{H}_{\text{off}}$ corresponds to an on-resonance experiment on our liquid test sample but applies also for sufficiently simple solid samples such as CaF_2, so long as we may disregard damping of the nuclear signal. The resulting x- and y-PSD outputs are shown schematically in Fig. 4-3a. Figure 4-3b shows oscilloscope traces of an actual experiment.[54]

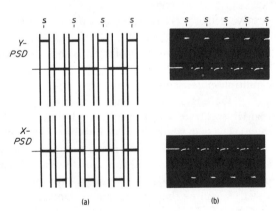

(a)   (b)

FIG. 4-3. Output signals of y- and x-PSDs for an on-resonance WAHUHA experiment on a liquid test sample. (a) Schematic, (b) actual experiment; $\tau = 6\ \mu\text{sec}$. Interruptions of the oscilloscope traces indicate pulses. Standard sampling points are indicated by s.

[54] Courtesy of U. Kohlschütter.

We now turn to the off-resonance cases, which are much more interesting if only for the reason that virtually all experiments are done off-resonance. The evolution of $\rho_T(t)$ is determined by Eqs. (4-16) and (4-17). For our test sample $\mathcal{H}_T^{int}(t) = \mathcal{H}_T^{off}(t)$. In the average Hamiltonian approximation we replace $\mathcal{H}_T^{off}(t)$ by its average:

$$\overline{\mathcal{H}_T^{off}(t)} = -\Delta\omega\tfrac{1}{3}(I_x+I_y+I_z) = -\Delta\omega_T \mathbf{I}\cdot\mathbf{n}, \tag{4-18}$$

where $\mathbf{n} = (1/\sqrt{3})(1,1,1)$, $|\mathbf{n}| = 1$, and $\Delta\omega_T = (1/\sqrt{3})\Delta\omega$. Hence,

$$\rho_T(t) \propto \exp[i\Delta\omega_T t\mathbf{I}\cdot\mathbf{n}]\, I_y \exp[-i\Delta\omega_T t\mathbf{I}\cdot\mathbf{n}]. \tag{4-19}$$

Equation (4-19) tells us that $\overline{\mathcal{H}_T^{off}(t)}$ imposes on the spin system a rotation about the toggling frame $(1,1,1)$ axis with angular velocity $\Delta\omega_T$. Alternatively, we may say that an effective field $\mathbf{B}_{eff,\,T} = +(1/\sqrt{3})\Delta B\mathbf{n}$ is operative in the toggling frame.

The ratio $\Delta\omega_T/\Delta\omega = 1/\sqrt{3}$ is called the "scaling factor." This scaling factor applies, evidently, also for genuine chemical shifts. We point out that the numerical value of the scaling factor is $1/\sqrt{3}$ only for an idealized WAHUHA experiment ($t_W \to 0$). Real WAHUHA and other real or idealized multiple-pulse sequences produce different scaling factors. In our line-narrowing experiments we always check the scaling factor experimentally, e.g., in the manner described in Mansfield and Haeberlen.[55]

In order to evaluate the NMR signals in the off-resonance cases it is convenient to introduce a "new" frame of reference (X, Y, Z), the Z axis of which points along $\mathbf{n} = (1/\sqrt{3})(1,1,1)$. The transformation formulas are (see Fig. 4-4)

$$I_X = (1/\sqrt{6})(I_x+I_y) - \sqrt{\tfrac{2}{3}}I_z,$$

$$I_Y = -(1/\sqrt{2})(I_x-I_y),$$

$$I_Z = (1/\sqrt{3})(I_x+I_y+I_z),$$

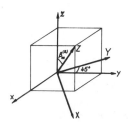

FIG. 4-4. Orientation of "new frame" (X, Y, Z) with respect to "old frame" (x, y, z).

[55] P. Mansfield and U. Haeberlen, Z. Naturforsch. A 28, 1081 (1973).

and

$$I_x = (1/\sqrt{6})I_X - (1/\sqrt{2})I_Y + (1/\sqrt{3})I_Z,$$
$$I_y = (1/\sqrt{6})I_X + (1/\sqrt{2})I_Y + (1/\sqrt{3})I_Z,$$
$$I_z = -\sqrt{\tfrac{2}{3}}I_X + (1/\sqrt{3})I_Z.$$

Expressed in the new frame $\rho_T(t)$ becomes

$$\rho_T(t) \propto \exp(i\,\Delta\omega_T\, tI_Z)\{(1/\sqrt{6})I_X + (1/\sqrt{2})I_Y + 1(/\sqrt{3})I_Z\}\exp(-i\,\Delta\omega_T\, tI_Z). \tag{4-20}$$

Now we can apply the formulas of Appendix A to obtain

$$\rho_T(t) \propto \frac{1}{\sqrt{6}}(I_X\cos\Delta\omega_T\, t - I_Y\sin\Delta\omega_T\, t)$$
$$+ \frac{1}{\sqrt{2}}(I_Y\cos\Delta\omega_T\, t + I_X\sin\Delta\omega_T\, t) + \frac{1}{\sqrt{3}}I_Z.$$

In the "old" toggling frame (x, y, z) we have

$$\rho_T(t) \propto \left[\frac{2}{3}I_y - \frac{1}{3}(I_x + I_z)\right]\cos\Delta\omega_T\, t$$
$$+ \frac{1}{\sqrt{3}}(I_x - I_z)\sin\Delta\omega_T\, t + \frac{1}{3}(I_x + I_y + I_z). \tag{4-21}$$

We are now ready to write expressions for the NMR signals after phase-sensitive detection by x- and y-PSDs. In the "observation windows" of the WAHUHA pulse train $\rho_R(t) = \rho_T(t)$ and the NMR signal produced by a y-PSD is proportional to

$$\mathrm{tr}\,\rho_R(t)I_y = \mathrm{tr}\,\rho_T(t)I_y \propto \tfrac{2}{3}\cos\Delta\omega_T\, t + \tfrac{1}{3} \qquad \text{(observation windows)}.$$

Recall $\mathrm{tr}\,I_x I_z = \mathrm{tr}\,I_x I_y = \mathrm{tr}\,I_y I_z = 0$.

In both types of "small" windows

$$\rho_R(t) = P_{-x}\rho_T(t)P_{-x}^{-1} = \exp[-\tfrac{1}{2}i\pi I_x]\rho_T(t)\exp[+\tfrac{1}{2}i\pi I_x]$$
$$\propto \left[\frac{2}{3}I_z - \frac{1}{3}(I_x - I_y)\right]\cos\Delta\omega_T\, t + \frac{1}{\sqrt{3}}(I_x + I_y)\sin\Delta\omega_T\, t + \frac{1}{3}(I_x + I_y + I_z). \tag{4-22}$$

The NMR signal produced by a y-PSD is proportional to

$$\mathrm{tr}\,\rho_R(t)I_y \propto \frac{1}{3}\cos\Delta\omega_T\, t + \frac{1}{\sqrt{3}}\sin\Delta\omega_T\, t - \frac{1}{3}$$
$$= \frac{2}{3}\cos\left(\Delta\omega_T\, t - \frac{\pi}{3}\right) - \frac{1}{3} \qquad \text{(small windows)}.$$

In the second type of "large" windows

$$\rho_R(t) = P_y P_{-x} \rho_T(t) P_{-x}^{-1} P_y^{-1}$$

$$= \exp(\tfrac{1}{2}i\pi I_y) \exp(-\tfrac{1}{2}i\pi I_x) \rho_T(t) \exp(\tfrac{1}{2}i\pi I_x) \exp(-\tfrac{1}{2}i\pi I_y)$$

$$\propto \left[-\frac{2}{3}I_x - \frac{1}{3}(I_z - I_y) \right] \cos \Delta\omega_T t$$

$$+ \frac{1}{\sqrt{3}}(I_y + I_z) \sin \Delta\omega_T t + \frac{1}{3}(-I_x - I_y + I_z).$$

Since the I_y operators are not affected by P_y, P_y^{-1}, the NMR signal produced by a y-PSD is the same as for the small windows.

An x-PSD produces NMR signals proportional to

$$\mathrm{tr}\,\rho_R I_x \propto -\frac{1}{3} \cos \Delta\omega_T t + \frac{1}{\sqrt{3}} \sin \Delta\omega_T t + \frac{1}{3}$$

$$= -\frac{2}{3} \cos\left(\Delta\omega_T t + \frac{\pi}{3} \right) + \frac{1}{3} \quad \text{(observation and small windows)},$$

$$\mathrm{tr}\,\rho_R I_x \propto -\frac{2}{3} \cos \Delta\omega_T t - \frac{1}{3} \quad \text{(other large windows)}.$$

We see that both x- and y-PSDs give just two types of signals consisting of oscillations and dc levels. We say that the oscillations "ride" on "pedestals." The initial amplitudes of the oscillations are $\frac{2}{3}$ of the initial amplitude of a free induction decay signal.

The pedestals are half as large as the amplitudes of the oscillations and have opposite signs for the two types of signals. The physical origin of the pedestals is obviously a nonzero component of the initial magnetization parallel to the effective field $\mathbf{B}_{\mathrm{eff},\,T}$ in the toggling frame. The pedestals can be avoided—and the useful signal amplitude increased by 50%—by applying properly chosen preparation pulses prior to the WAHUHA pulse train proper. A 45° y pulse immediately preceding the first x pulse (which we have termed preparation pulse so far) does the job.

Relaxation processes cause a decay of both the oscillations and the pedestals. We shall come back to that point in Chapter V, Section C. Here we mention only that the decay of the pedestals involves spin lattice relaxation processes, whereas the decay of the oscillations does not, at least not necessarily. Therefore, the oscillations usually decay much faster than the pedestals.

It is easy to see that three types of signals appear as outputs of PSDs, the reference phases of which are anywhere between x and y. To keep matters as simple as possible one usually avoids such phase settings.

Figure 4-5a,b shows y signals "below" and "above" resonance. The time

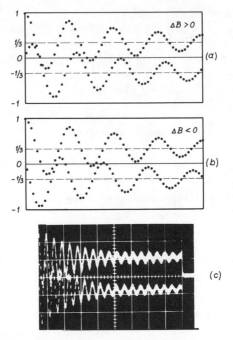

FIG. 4-5. NMR signals during WAHUHA pulse train after y-phase-sensitive detection. (a) Above, (b) below resonance; the time scale is much larger than in Fig. 4-3. (c) Actual experiment below resonance on CaF_2. Time scale: 1 msec/division; cycle time $t_c = 24$ μsec. Equal amplitudes of the oscillations, up- as well as downfield, and correct pedestals are important criteria for the correct alignment of the sequence.

scale is much larger than in Fig. 4-3. A decay of the oscillations, but not of the pedestals has been taken into account. Figure 4-5c shows oscilloscope traces of an actual experiment on CaF_2.

So far our treatment of averaging in r and spin spaces was restricted. Only time averages were taken into account. Also, we restricted ourselves to an idealized WAHUHA pulse sequence insofar as widths of rf pulses were concerned. The average Hamiltonian theory[51] to be discussed now will provide us with a tool for a more complete treatment of averaging processes.

D. Average Hamiltonian Theory

1. Interaction Representations Defined by Explicitly Time-Dependent Hamiltonians

We consider a system with a total Hamiltonian

$$\mathscr{H} = \mathscr{H}_1(t) + \mathscr{H}^{int}, \tag{4-23}$$

where $\mathcal{H}_1(t)$ is explicitly dependent upon time and $\mathcal{H}^{\text{int}}$ is the "internal" or "interesting" part of the Hamiltonian. In analogy to Eq. (4-10) we make the following *ansatz* for the density matrix $\rho(t)$:

$$\rho(t) = U_1(t)\tilde{\rho}(t) U_1^{-1}(t), \tag{4-24}$$

where

$$U_1(t) = T\exp\left[-i\int_0^t \mathcal{H}_1(t')\,dt'\right]. \tag{4-25}$$

T is the Dyson time-ordering operator (cf. Appendix B). On expanding the exponential in Eq. (4-25) products of the type $\cdots\mathcal{H}_1(t''')\mathcal{H}_1(t'')\mathcal{H}_1(t')\cdots$ are obtained. T ensures that operators that act at earlier times on $\tilde{\rho}(t)$ are always at the right of operators that act at later times on $\tilde{\rho}(t)$. The opposite order of operators is obviously needed in $U_1^{-1}(t)$.

$\tilde{\rho}(t)$ evolves according to

$$\frac{d\tilde{\rho}}{dt} = -i[\tilde{\mathcal{H}}^{\text{int}}(t), \tilde{\rho}(t)]^1, \tag{4-26}$$

where

$$\tilde{\mathcal{H}}^{\text{int}}(t) = U_1^{-1}(t)\mathcal{H}^{\text{int}}U_1(t). \tag{4-27}$$

The formal solution of Eq. (4-26) is

$$\tilde{\rho}(t) = U_{\text{int}}(t)\tilde{\rho}(0) U_{\text{int}}^{-1}(t), \tag{4-28}$$

where

$$U_{\text{int}}(t) = T\exp\left[-i\int_0^t \tilde{\mathcal{H}}^{\text{int}}(t')\,dt'\right]. \tag{4-29}$$

Combining Eqs. (4-24) and (4-28) yields

$$\rho(t) = U(t)\rho(0) U^{-1}(t), \tag{4-30}$$

where

$$U(t) = U_1(t) U_{\text{int}}(t). \tag{4-31}$$

2. CYCLIC INTERACTIONS

An interaction $\mathcal{H}_1(t)$ of a system is said to be cyclic if $\mathcal{H}_1(t)$ is periodic, i.e., if

$$\mathcal{H}_1(t+t_{\text{h}}) = \mathcal{H}_1(t), \tag{4-32}$$

and if, in addition,

$$U_1(t) = T\exp\left[-i\int_0^t \mathcal{H}_1(t')\,dt'\right]$$

is also periodic, i.e., if

$$U_1(t) = U_1(t+t_{\text{c}}). \tag{4-33}$$

If conditions (4-32) and (4-33) are satisfied[56] the period t_c of $U_1(t)$ will be an integer multiple of the period t_h of $\mathscr{H}_1$. t_c is called cycle time. Intervals of $\mathscr{H}_1(t)$ of duration t_c are called cycles.

Since $U_1(0) = 1$ an immediate consequence of Eq. (4-33) is

$$U_1(Nt_c) = T\exp\left[-i\int_0^{Nt_c}\mathscr{H}_1(t')\,dt'\right] = 1 \qquad \text{for} \quad N \text{ integer.} \qquad (4\text{-}34)$$

We note in passing that Eq. (4-34) [not Eq. (4-33)!] may be satisfied for shorter time intervals than t_c. When this occurs one speaks of subcycles of duration t_s.

We give now a few examples of cyclic spin interactions.

(1) It is obvious from Fig. 4-2 that the WAHUHA sequence is cyclic. It was, in fact, this pulse sequence that led to the introduction of the notion of a cycle. The cycle time t_c is equal to the period 6τ of the pulse sequence. Note that the cyclic property of the WAHUHA sequence is not destroyed by finite pulse widths t_w. Note also that a choice of $U_{rf}(t)$ other than ours—e.g., no pulse treated separately as a preparation pulse—would reveal the existence of subcycles of duration 3τ.

(2) Another example for a cyclic pulse sequence is the Carr–Purcell sequence[58] with or without the Gill–Meiboom modification[59] of the preparation pulse. The cycle time t_c is twice the period of the pulse sequence.

(3) A cyclic interaction of a spin system that does not originate in a pulse sequence is the Zeeman interaction of a system of like spins with a constant, homogeneous applied magnetic field. The period of $\mathscr{H}_1 = \mathscr{H}_z$ is degenerate, i.e., infinitely small. The cycle time, however, is well defined: it is equal to the period of the Larmor precession of the spins.

The same applies for constant fields in the rotating frame. Examples are fields used in the spin-locking[60] and Lee–Goldburg[61] experiments.

An example where the periodicity of $\mathscr{H}_1(t)$ does not entail a periodicity of $U_1(t)$ is the interaction of a spin system with a periodic sequence of on-resonance rf pulses the flip angle β of which is not a rational fraction of 2π.

Hamiltonians with a cyclic term $\mathscr{H}_1(t)$ are the realm of the average Hamiltonian theory proper, which we shall now outline. It exploits the fact that $U_1(t)$ transfers its periodicity to $\tilde{\mathscr{H}}^{int}(t)$ [cf. Eq. (4-27)]:

$$\tilde{\mathscr{H}}^{int}(t+Nt_c) = \tilde{\mathscr{H}}^{int}(t) \qquad \text{for} \quad N \text{ integer.} \qquad (4\text{-}35)$$

[56] We do not inquire here into the general conditions under which cyclic interactions are possible at all. This question is raised, and answered in part, by Salzman.[57]

[57] W. R. Salzman, *Phys. Rev. A* **10**, 461 (1974).

[58] H. Y. Carr and E. M. Purcell, *Phys. Rev.* **94**, 630 (1954).

[59] D. Gill and S. Meiboom, *Rev. Sci. Instrum.* **29**, 688 (1958).

[60] A. G. Redfield, *Phys. Rev.* **98**, 1787 (1955).

[61] M. Lee and W. I. Goldburg, *Phys. Rev.* **140**, A1261 (1965).

A consequence of Eq. (4-35) is

$$U_{\text{int}}(Nt_c) = T\exp\left[-i\int_0^{Nt_c} \mathcal{H}^{\text{int}}(t')\,dt'\right] = \left\{T\exp\left[-i\int_0^{t_c} \mathcal{H}^{\text{int}}(t')\,dt'\right]\right\}^N$$

$$= [U_{\text{int}}(t_c)]^N. \tag{4-36}$$

Combining this result with Eqs. (4-31) and (4-34) gives

$$\rho(Nt_c) = [U_{\text{int}}(t_c)]^N \rho(0)[U_{\text{int}}^{-1}(t_c)]^N. \tag{4-37}$$

The content of Eq. (4-37) can be stated as follows. In order to describe the state of the system at any integer multiple of the cycle time t_c it is sufficient to know its short-time evolution over one cycle. The one-cycle propagator $U_{\text{int}}(t_c)$ is simply raised to the Nth power.

3. MAGNUS EXPANSION

To obtain the result in the form of a single exponential it is convenient to apply the Magnus formula[62] to $U_{\text{int}}(t_c)$:

$$U_{\text{int}}(t_c) = T\exp\left[-i\int_0^{t_c} \mathcal{H}^{\text{int}}(t')\,dt'\right] \qquad \text{(definition)}$$

$$= \exp[-iFt_c] \qquad \text{(ansatz)}$$

$$= \exp[-i(\bar{\mathcal{H}} + \bar{\mathcal{H}}^{(1)} + \bar{\mathcal{H}}^{(2)} + \cdots)t_c] \qquad \text{(expansion)}, \tag{4-38}$$

where

$$\bar{\mathcal{H}} = \frac{1}{t_c}\int_0^{t_c} \mathcal{H}^{\text{int}}(t)\,dt, \tag{4-39}$$

$$\bar{\mathcal{H}}^{(1)} = \frac{-i}{2t_c}\int_0^{t_c} dt_2 \int_0^{t_2} dt_1 \,[\mathcal{H}^{\text{int}}(t_2), \mathcal{H}^{\text{int}}(t_1)], \tag{4-40}$$

$$\bar{\mathcal{H}}^{(2)} = \frac{-1}{6t_c}\int_0^{t_c} dt_3 \int_0^{t_3} dt_2 \int_0^{t_2} dt_1 \,\{[\mathcal{H}^{\text{int}}(t_3), [\mathcal{H}^{\text{int}}(t_2), \mathcal{H}^{\text{int}}(t_1)]]$$

$$+ [\mathcal{H}^{\text{int}}(t_1), [\mathcal{H}^{\text{int}}(t_2), \mathcal{H}^{\text{int}}(t_3)]]\}, \tag{4-41}$$

etc. A number of rigorous, mostly algebraic derivations of these expressions have been given in the mathematical literature.[62,63] Nevertheless, we indicate one in Appendix B that follows closely the line of our later applications and should therefore be instructive.

Each of the terms of the Magnus expansion is Hermitian.[64] Truncation of

[62] W. Magnus, *Commun. Pure Appl. Math.* **7**, 649 (1954).
[63] R. M. Wilcox, *J. Math. Phys.* **8**, 962 (1967).
[64] P. Pechukas and F. C. Light, *J. Chem. Phys.* **44**, 3897 (1966).

the expansion at any place therefore always leaves us with a "legitimate" Hamiltonian.

If $\mathscr{H}^{int}(t)$ commuted with itself at all times, $\mathscr{H}$ (which we shall continue to call "average Hamiltonian") alone would suffice to describe the time development of the system. Sometimes this is the case; more often, however, $\mathscr{H}^{int}(t)$ does not behave so simply and the higher order average Hamiltonians $\mathscr{H}^{(n)}$ must be included. In a sense they can be thought of as quantum corrections to a "classical" theory.

Combining Eqs. (4-37) and (4-38) and recalling that F is a time-independent operator, we get

$$\rho(Nt_c) = \exp[-iFNt_c]\,\rho(0)\exp[+iFNt_c], \qquad (4\text{-}42)$$

or, if it is tacitly understood that the real time t is restricted to integer multiples of t_c,

$$\rho(t) = \exp[-iFt]\,\rho(0)\exp[+iFt]. \qquad (4\text{-}43)$$

This is the central result of the average Hamiltonian theory. It states that a system subjected to cyclic (mostly external) forces, the inspection of which is restricted to integer multiples of the cycle time t_c, behaves as if it were developing according to a constant Hamiltonian F. The lowest order approximation of F is $\mathscr{H}$, which is the time average of $\mathscr{H}^{int}(t)$, not of $\mathscr{H}^{int}$. The Magnus expansion tells us how $\mathscr{H}$ is to be supplemented when the lowest order approximation does not provide us with sufficiently accurate results.

When $\mathscr{H}^{int}$, or any one term of it, possesses a periodic time dependence of its own—e.g., through a time-dependent parameter—the Magnus expansion applies, of course, also in the absence of any cyclic interaction $\mathscr{H}_1(t)$. In such cases $\mathscr{H}$ is truly the average of $\mathscr{H}^{int}(t)$. The Magnus expansion provides, in fact, the basis for the general rule stated in Section A, on which we have drawn so extensively throughout this text.

To get a rough idea of the importance of the correction terms $\mathscr{H}^{(1)}$, $\mathscr{H}^{(2)}$, etc., let us consider first the particularly unfavorable case in which $\mathscr{H}$, $\mathscr{H}^{(1)}$, etc., exhibit no specific averaging. The strength of $\mathscr{H}$ will then be of the same order of magnitude as the strength of $\mathscr{H}^{int}$ itself. Specifically let us think of $\mathscr{H}^{int}$ as representing the direct and indirect spin–spin interactions of a rigid solid. A measure for the strength of $\mathscr{H}^{int}$ is $\langle\Delta\omega^2\rangle^{1/2}$, the root of the second moment of the resonance line. Assuming gaussian lineshapes the FID will have decayed to $e^{-1/2}$ in a time $\langle\Delta\omega^2\rangle^{-1/2}$.

In an unspecific multiple-pulse experiment we expect from the form of Eqs. (4-39)–(4-41) decay rates associated with the various terms $\mathscr{H}^{(n)}$ of the order of $\langle\Delta\omega^2\rangle^{1/2}(t_c\langle\Delta\omega^2\rangle^{1/2})^n$, which become rapidly small with increasing order n if $t_c\langle\Delta\omega^2\rangle^{1/2}$ is appreciably smaller than unity. Recall that in a typical case t_c can be chosen by the experimenter, although often there are practical limitations preventing him from making t_c as short as he would like.

A more favorable situation exists if we can arrange—as we often can—that $\mathscr{H} = \mathscr{H}^{(1)} = \cdots = \mathscr{H}^{(n-1)} = 0$. We then find (see Appendix B)

$$\mathscr{H}^{(n)} = \frac{(-i)^n}{t_c} \int_0^{t_c} dt_{n+1} \int_0^{t_{n+1}} dt_n \cdots \int_0^{t_2} dt_1 \, \mathscr{H}^{int}(t_{n+1}) \, \mathscr{H}^{int}(t_n) \cdots \mathscr{H}^{int}(t_1),$$

(4-44)

and since the volume of integration is $(t_c)^{n+1}/(n+1)!$ the first nonvanishing correction term $\mathscr{H}^{(n)}$ will lead to a decay rate not larger than

$$[1/(n+1)!] \langle \Delta\omega^2 \rangle^{1/2} (t_c \langle \Delta\omega^2 \rangle^{1/2})^n.$$

4. SYMMETRIC AND ANTISYMMETRIC CYCLES

Another special case that can often be realized in practice is

$$\mathscr{H}^{int}(t) = \mathscr{H}^{int}(t_c - t).$$

Cycles that generate this property of $\mathscr{H}^{int}(t)$ are said to be symmetric. Mansfield[53,65] first realized that $\mathscr{H}^{(1)}$ vanishes for symmetric cycles. Wang and Ramshaw[66] have given an elegant proof (which we essentially reproduce in Appendix B) for the fact that all correction terms $\mathscr{H}^{(k)}$, with k odd, vanish for symmetric cycles.

As a kind of curiosity we add that $\mathscr{H}$ itself vanishes together with all correction terms, even and odd, if $\mathscr{H}^{int}(t)$ is antisymmetric, i.e., if $\mathscr{H}^{int}(t) = -\mathscr{H}^{int}(t_c - t)$.

In the remaining sections of this chapter we shall apply the average Hamiltonian theory to analyze a selection of experiments and phenomena in NMR. When the lowest order, the average Hamiltonian, approximation turns out to be inadequate we shall take into account correction terms $\mathscr{H}^{(n)}$, $n \geqslant 1$, of the Magnus expansion. Taking into account $\mathscr{H}^{(1)}$ is equivalent to standard second-order perturbation theory.

In Chapter V we shall see how the average Hamiltonian theory can help to design more and more powerful line-narrowing multiple-pulse sequences.

E. Stochastic Averaging. Comparison with Coherent Averaging

So far we have applied the Magnus expansion to Hamiltonians that are periodic in time [see Eq. (4-35)]. Now let us study its application to stochastic Hamiltonians. This will lead us to the master equation of the density matrix ρ.

---

[65] He failed, however, to recognize that the WAHUHA cycle *is* symmetric if one chooses the "sampling points" in the middle of the windows that we termed "observation windows," as one does in actual experiments.

[66] C. H. Wang and J. D. Ramshaw, *Phys. Rev. B* **6**, 3253 (1972).

Let the system to be considered be characterized at time $t = 0$ by the density matrix ρ_0 and let it be subjected to the stochastic Hamiltonian $\mathscr{H}^{st}(t)$, the correlation time of which is τ_c. After a time t_s (s stands for sampling) the density matrix will be

$$\rho(t_s) = U_{st}(t_s)\,\rho_0\,U_{st}^{-1}(t_s).$$

Nothing prevents us from expressing $U_{st}(t_s) = T \exp[-i\int_0^{t_s}\mathscr{H}^{st}(t')\,dt']$ by

$$U_{st}(t_s) = \exp[-it_s(\bar{\mathscr{H}} + \bar{\mathscr{H}}^{(1)} + \cdots)]. \tag{4-45}$$

$\bar{\mathscr{H}}$ and $\bar{\mathscr{H}}^{(1)}$ are defined by Eqs. (4-39) and (4-40) with $\mathscr{H}^{int}(t)$ and t_c replaced by $\mathscr{H}^{st}(t)$ and t_s, respectively. Expanding the exponential in Eq. (4-45) leads to

$$\rho(t_s) = \rho_0 - it_s[(\bar{\mathscr{H}} + \bar{\mathscr{H}}^{(1)} + \cdots), \rho_0] - \tfrac{1}{2}t_s^2[[\rho_0, \bar{\mathscr{H}}], \bar{\mathscr{H}}] + \cdots$$

$$= \rho_0 - i\int_0^{t_s}[\mathscr{H}^{st}(t'), \rho_0]\,dt'$$

$$\underbrace{- \tfrac{1}{2}\int_0^{t_s}dt''\int_0^{t''}dt'[[\mathscr{H}^{st}(t''), \mathscr{H}^{st}(t')], \rho_0]}_{I_1}$$

$$\underbrace{- \tfrac{1}{2}\int_0^{t_s}dt''\int_0^{t_s}dt'[[\rho_0, \mathscr{H}^{st}(t'')], \mathscr{H}^{st}(t')]}_{I_2} + \cdots \tag{4-46}$$

Now concentrate for a while on I_1 and I_2. The domains of integration of these integrals are different. In order to make them equal we rewrite I_2:

$$I_2 = -\tfrac{1}{2}\int_0^{t_s}dt''\int_0^{t''}dt'\{[[\rho_0, \mathscr{H}^{st}(t'')], \mathscr{H}^{st}(t')] + [[\rho_0, \mathscr{H}^{st}(t')], \mathscr{H}^{st}(t'')]\}. \tag{4-47}$$

Combining I_1 and I_2 results in

$$I_1 + I_2 = -\int_0^{t_s}dt''\int_0^{t''}dt'[[\rho_0, \mathscr{H}^{st}(t')], \mathscr{H}^{st}(t'')].$$

The domain of integration is shown in Fig. 4-6 by the shaded area.

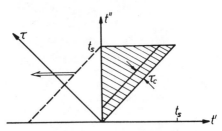

FIG. 4-6. Domain of integration of $I_1 + I_2$.

We proceed by introducing $\tau = t'' - t'$ as a new variable of integration. This leads to

$$I_1 + I_2 = -\int_0^{t_s} dt'' \int_0^{t''} d\tau \, [[\rho_0, \mathscr{H}^{st}(t'' - \tau)], \mathscr{H}^{st}(t'')].$$

Now we choose $t_s \gg \tau_c$. This is, in principle, a task for the experimenter. Because of the stochastic nature of $\mathscr{H}^{st}(t)$ the only area whose contributions to the integral do not cancel is a narrow strip along the line $\tau = 0$. The breadth of the strip is, of course, not well defined, but in any case it is of the order of a few τ_c. It is disturbing that t'', the variable of the outer integral appears as an integration limit of the inner one. To get rid of it we extend the domain of integration—without altering appreciably the value of the integral—to the dashed line in Fig. 4-6, which corresponds to $\tau = t_s$, and eventually even to $\tau \to \infty$. This leads to

$$I_1 + I_2 = -\int_0^{t_s} dt'' \int_0^{\infty} d\tau \, [[\rho_0, \mathscr{H}^{st}(t'' - \tau)], \mathscr{H}^{st}(t'')],$$

or

$$\rho(t_s) = \rho_0 - i \int_0^{t_s} [\mathscr{H}^{st}(t''), \rho_0] \, dt''$$
$$- \int_0^{t_s} dt'' \int_0^{\infty} d\tau \, [[\rho_0, \mathscr{H}^{st}(t'' - \tau)], \mathscr{H}^{st}(t'')] + \cdots. \quad (4\text{-}48a)$$

We come to a very delicate point now. In order to pin it down, drop for the moment the index s from t_s in Eq. (4-48a). Differentiating with respect to time seems to be a trivial matter: just drop ρ_0 and the symbols $\int_0^{t_s} dt''$, and replace t'' by t in the integrands. The result, however, would be wrong and therefore we avoid writing it. What is the trouble? The careless differentiation that we indicated implied the limiting process $t \to 0$. This destroys, however, the basis on which rests the extension of the domain of integration discussed above. One way of circumventing this difficulty is the following. First we rewrite Eq. (4-48a):

$$\frac{\rho(t_s) - \rho_0}{t_s} = -i\frac{1}{t_s} \int_0^{t_s} [\mathscr{H}^{st}(t''), \rho_0] \, dt''$$
$$- \frac{1}{t_s} \int_0^{t_s} dt'' \int_0^{\infty} d\tau \, [[\rho_0, \mathscr{H}^{st}(t'' - \tau)], \mathscr{H}^{st}(t'')] + \cdots.$$
$$(4\text{-}48b)$$

Then we stress again that the expansion from which Eq. (4-48b) was derived restricts its range of validity to small values of t_s. On the other hand, the extension of the domain of integration that is incorporated in Eq. (4-48b) restricts its range of validity to large values of t_s. To reconcile these two contradicting requirements remember what small and large mean.

Small means small on the time scale characteristic of the large-scale variations of $\rho(t)$, which means in practice small compared with the nuclear spin relaxation times T_1, T_2. [We contrast large-scale variations of $\rho(t)$ against fast, small-amplitude and phase fluctuations of $\rho(t)$.]

Large means, on the other hand, large compared with τ_c.

If $\tau_c \ll T_1, T_2$ it is possible to find a time T such that

$$\lim_{t_s \to T} \left\{ \frac{\rho(t_s) - \rho_0}{t_s} \right\} \approx \lim_{t_s \to 0} \left\{ \frac{\rho(t_s) - \rho_0}{t_s} \right\} \equiv \dot{\rho}/_0, \qquad (4\text{-}49)$$

and simultaneously

$$\lim_{t_s \to T} \left\{ \frac{1}{t_s} \int_0^{t_s} (\cdots)\, dt'' \right\} \approx \lim_{t_s \to \infty} \left\{ \frac{1}{t_s} \int_0^{t_s} (\cdots)\, dt'' \right\} \equiv \langle(\cdots)\rangle_{\text{average over } t''}, \qquad (4\text{-}50)$$

where $(\cdots)$ means either $[\mathscr{H}^{\text{st}}(t''), \rho_0]$ or $\int_0^\infty d\tau\, [[\rho_0, \mathscr{H}^{\text{st}}(t'' - \tau)], \mathscr{H}^{\text{st}}(t'')]$. According to the ergodic hypothesis the average over t may be replaced by an average over a large ensemble. One usually defines $\mathscr{H}^{\text{st}}(t)$ such that

$$\langle \mathscr{H}^{\text{st}}(t) \rangle_{\text{av}} = 0 \quad \text{which implies} \quad \langle [\mathscr{H}^{\text{st}}(t), \rho_0] \rangle_{\text{av}} = 0. \quad (4\text{-}51)$$

Combining Eqs. (4-48)–(4-51) gives an equation for $\dot{\rho}/_0$.

There is nothing special about our particular choice of the initial time $t = 0$; therefore, we claim that the resulting equation holds for all times t if we replace ρ_0 and $\dot{\rho}/_0$ by $\rho(t)$ and $\dot{\rho}(t)$. What results is the master equation of the density matrix ρ:

$$d\rho(t)/dt = -\int_0^\infty d\tau \langle [[\rho(t), \mathscr{H}^{\text{st}}(t - \tau)], \mathscr{H}^{\text{st}}(t)] \rangle_{\text{av}}. \qquad (4\text{-}52)$$

If the Hamiltonian of the system is periodic instead of stochastic we get instead of Eq. (4-52) the derivative of Eq. (4-43),

$$d\rho_c(t)/dt = -i[F, \rho_c(t)] = -i[(\bar{\mathscr{H}} + \bar{\mathscr{H}}^{(1)} + \bar{\mathscr{H}}^{(2)} + \cdots), \rho_c(t)], \qquad (4\text{-}53)$$

which is the counterpart of the master equation for coherent averaging. (The label c stands for coherent.)

We compare now stochastic with coherent averaging. Both cases are similar insofar as differentiating and differential quotients are concerned: The left-hand of Eqs. (4-52) and (4-53) are not differential quotients in the strict mathematical sense of the word. In both cases the "differential" dt must stay substantially larger than τ_c and t_c, respectively. Nevertheless, these "differential quotients" do have a practical meaning—even an immensely important one—provided the large-scale variations of the density matrix occur on a time scale much slower than the one established by τ_c or t_c.

The first dissimilarity to note is that in the coherent case t and any Δt must be restricted to integer multiples of t_c, whereas in the stochastic case t and Δt may be varied smoothly as long as they remain large compared with τ_c. This last condition, however, renders this dissimilarity rather immaterial.

The second difference concerns the very nature of Eqs. (4-52) and (4-53). The master equation describes an irreversible [Eq. (4-53)], on the other hand, a quantum mechanical, and therefore, in principle, a reversible, evolution of a (single- or many-particle) system.

The external signs of this difference are

(a) The absence and presence, respectively, of the factor i in Eqs. (4-52) and (4-53).

(b) The brackets indicating a statistical average on the right-hand side of the master equation. Such a statistical average is missing in Eq. (4-53).

There is, of course, a difference with respect to selectiveness: coherent averaging typically is selective, stochastic averaging typically is *not*.

The last difference to which we would like to draw the reader's attention concerns the efficiency of the averaging. It is a common practice to correlate the stochastically narrowed linewidth $\delta\omega$ with the second moment $\langle\Delta\omega^2\rangle$ of the rigid lattice by the formula

$$\delta\omega^2 = \langle\Delta\omega^2\rangle (2/\pi) \arctan[\alpha\,\delta\omega\tau_c] \qquad (4\text{-}54)$$

(Abragam,[9] p. 456), where α is a rather ill-defined factor of order unity. For $\alpha = 1$ and, e.g., $\tau_c\langle\Delta\omega^2\rangle^{1/2} = 1$, 0.5, 0.2, Eq. (4-54) predicts 1.73, 3.17, and 7.86, respectively, for the linewidth reduction factor $(\langle\Delta\omega^2\rangle/\delta\omega^2)^{1/2}$.

Coherent averaging by, e.g., the WAHUHA sequence produces much larger reduction factors for comparable values of $t_c\langle\Delta\omega^2\rangle^{1/2}$. Our early multiple-pulse experiments[46, 51] with cycle times $t_c = 24$ μsec demonstrated for CaF_2 linewidth reduction factors in excess of 100. For CaF_2 oriented with its (111) axis along $\mathbf{B}_{st}$, $\langle\Delta\omega^2\rangle^{1/2}$ is 3.85×10^4 sec^{-1}; this means for the product $t_c\langle\Delta\omega^2\rangle^{1/2}$ that it was very close to unity (0.92). In more recent experiments with still about the same values for t_c, reduction factors much larger than 100 have been achieved.[67]

To get a feeling for the reason of the superior efficiency of coherent averaging recall that in a WAHUHA experiment, e.g., the average of $\mathscr{H}^D(t)$ over t_c is zero, whereas for stochastic averaging $\mathscr{H}^{st}(t)$ does not average out in the time span τ_c; it does so only when the averaging time approaches infinity.

There would be no practical chance for high-resolution NMR in solids by multiple-pulse techniques if for achieving a certain line-narrowing effect the cycle time t_c had to be as short as the correlation time τ_c of stochastic

[67] W.-K. Rhim, D. D. Elleman, and R. W. Vaughan, *J. Chem. Phys.* **58**, 1772 (1973).

motions has to be. One of the great merits of Waugh is to have had faith in coherent averaging prior to experimental verification of its feasability and prior to detailed theoretical analyses of multiple-pulse sequences.

F. Further Examples of Averaging in Spin Space

One purpose of this section is to scan to some extent the range of phenomena where averaging in spin space plays an important role; another is to demonstrate with some fairly simple cases the usefulness of the Magnus expansion.

1. THE CARR–PURCELL (CP) SEQUENCE[58]

The pulse representation of this sequence is

$$P_\alpha^{90} \underbrace{- \tau - P_y^{180} - \tau - \tau - P_y^{180} - \tau - \tau - P_y^{180}}_{\text{cycle}} - \tau - \cdots$$

The sequence is cyclic, the cycle consisting of two 180° pulses. The cycle time t_c is 4τ. In the original version of the sequence $\alpha = y$ was chosen; in the later Gill–Meiboom modification[59] α was shifted to $+x$ or $-x$. The purpose of the CP sequence is to suppress the applied-field inhomogeneity term

$$\mathscr{H}^{\text{inh}} = -\gamma_n \sum_i \Delta B_0{}^i I_z{}^i = -\sum_i \Delta\omega^i I_z{}^i \tag{4-55}$$

from the effective Hamiltonian. $\Delta B_0{}^i$ is the deviation at the site of the ith nucleus of the z component of $\mathbf{B}_{\text{st}}(\mathbf{r})$ from its mean value. We point out that $\mathscr{H}^{\text{inh}}$ describes a truncated Hamiltonian in the sense of Section C,1. The interaction representation U_{rf}, which removes the pulses from the effective Hamiltonian, results in

$$\tilde{\mathscr{H}}^{\text{inh}}(t) = -\sum_i \Delta\omega^i \tilde{I}_z{}^i(t), \tag{4-56}$$

where for δ pulses $\tilde{I}_z{}^i(t)$ toggles between $+I_z{}^i$ and $-I_z{}^i$. As a result the average of $\tilde{\mathscr{H}}^{\text{inh}}(t)$ vanishes. If the width t_w of the 180° pulses is small enough to be neglected $\tilde{\mathscr{H}}^{\text{inh}}(t)$ commutes with itself at all times. It follows that all correction terms $\tilde{\mathscr{H}}^{(n)}$ ($n \geqslant 1$) of the Magnus expansion vanish. The inhomogeneity term is therefore suppressed completely by the CP sequence. The size of τ is of no importance.[68]

The correction terms $\tilde{\mathscr{H}}^{(n)}$ ($n \geqslant 1$) do not vanish if the necessarily finite width of the pulses is taken into account. It is instructive to consider these

[68] Note, however, that we *assumed* $\Delta B_0{}^i$ to be time independent. In fact, in many applications of the CP sequence—measurements of self-diffusion constants—$\Delta B_0{}^i$ is *not* time independent.

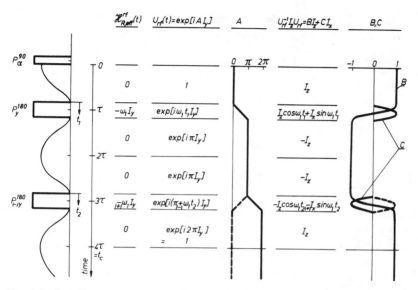

FIG. 4-7. Carr–Purcell sequence; time dependences of quantities of importance. Signs in parentheses, and dashed lines are for modification (a) of the sequence (see text).

correction terms in this very simple case, although the underlying problem can be treated explicitly—even classically,[69] since only single-spin operators are involved. We shall outline the discussion of the CP sequence with reference to Fig. 4-7.

The cycles are defined as the intervals from $t = 4N\tau$ to $t = (N+1)4\tau$, N integer. Accordingly, we imply that the even-numbered echos are observed. Obviously a finite width of the pulses does not cause $\mathcal{H}^{inh}$ to become different from zero (last column of Fig. 4-7). It is also obvious that $\mathcal{H}^{inh}(t)$ is not symmetric about the center of the cycle. As a result

$$\overline{\mathcal{H}}_{inh}^{(1)} = \frac{1}{\pi} t_W [1 - (t_W/t_c)] \sum_i (\Delta\omega^i)^2 I_y^i \neq 0. \qquad (4\text{-}57)$$

There are several ways to render $\mathcal{H}^{inh}(t)$ symmetric about the center of the cycle and thereby to kill $\overline{\mathcal{H}}_{inh}^{(1)}$, $\overline{\mathcal{H}}_{inh}^{(3)}$, etc.

(a)   One reverses the rf phase of alternating pulses. This is indicated in

[69] It has been treated classically, but with a lot of trigonometry, algebra, and numerical calculations by Hausser and Noack.[70] The average Hamiltonian formalism allows an analysis that is at least equivalent but that requires practically no calculations and leads, in addition, almost automatically to compensation schemes.

[70] R. Hausser and F. Noack, Z. Naturforsch. A 19, 1521 (1964).

Fig. 4-7 by the signs in parentheses and by the dashed curves A and C, which indeed display the desired symmetry about $t = 2\tau$. However,

$$\bar{\mathscr{H}}_{\text{inh}} = -\frac{4}{\pi} \frac{t_{\text{W}}}{t_{\text{c}}} \sum_{i} \Delta\omega^i I_x{}^i \neq 0. \tag{4-58}$$

We comment on this result below.

(b) One reverses the rf phase of alternating pairs of pulses. The cycle time becomes 8τ instead of 4τ. Only every fourth echo is observed. $\mathscr{H}^{\text{inh}}(t)$ becomes symmetric about the centers of the cycles. Both $\bar{\mathscr{H}}_{\text{inh}}$ and $\bar{\mathscr{H}}_{\text{inh}}^{(\text{odd})}$ vanish.

Constructing larger cycles out of smaller "basic" ones with reversed phases is a standard technique in high-resolution NMR in solids to get improved resolution. It seems, however, that modification (b) of the CP sequence has not yet been tried in actual experiments. The reason is probably that the Gill–Meiboom modification ($\alpha = \pm x$) achieves (almost) the same effect, though in a totally different way:

A P_x^{90} preparation pulse secures in liquid samples $\rho_R(0) \propto I_y$, which commutes with $\bar{\mathscr{H}}_{\text{inh}}^{(1)}$ as given by Eq. (4-57). $\bar{\mathscr{H}}_{\text{inh}}^{(1)}$ thus does not contribute to the initial decay of the observed echos. We cautiously stated "initial decay" since $\bar{\mathscr{H}}_{\text{inh}}^{(2)}$—which turns out to be proportional to I_z—will eventually rotate the magnetization away from the rotating frame y axis and then $\bar{\mathscr{H}}_{\text{inh}}^{(1)}$ can become an effective damping mechanism for the echo envelope.

Similarly, for modification (a) the resulting average inhomogeneity Hamiltonian [Eq. 4-58)] commutes with ρ_R (initial) provided a P_y^{90} preparation pulse is used.

Flip Angle Errors, rf Inhomogeneity

As a foretaste of the host of pulse errors we need to consider (and to fight) in high-resolution NMR in solids let us study what happens in a CP train if the 180° pulses are not perfect, in the sense that $\omega_1 t_{\text{w}} = \pi + \varepsilon$, $\varepsilon \ll \pi$, but $\neq 0$.

Because of the unavoidable inhomogeneity of the rf field it is impossible to adjust t_{w} such that $\varepsilon = 0$ for all parts of the sample. A discussion of flip angle errors is therefore automatically also a discussion of the effects of the rf inhomogeneity.

The first thing to note is that flip angle errors destroy the cyclic property of the CP sequence in its original version. This seems to be disastrous, at least for our formalism, but we can cure the trouble easily in the following way: we partition the rf field amplitude (as we did before with $\mathbf{B}_{\text{st}}$) into

$$B_1 = B_1{}^0 + \Delta B_1$$

such that

$$|\tfrac{1}{2}\gamma_{\text{n}} B_1{}^0 t_{\text{w}}| = \omega_1{}^0 t_{\text{w}} = \pi \quad \text{and} \quad |\tfrac{1}{2}\gamma_{\text{n}} \Delta B_1 t_{\text{w}}| = \Delta\omega_1 t_{\text{w}} = \varepsilon.$$

Only the "ideal" part $B_1{}^0$ of the rf field is removed from the effective Hamiltonian by the interaction representation U_{rf}. The remainder is kept as a pulse error Hamiltonian $\mathcal{H}^{err}(t)$. Only spin operators $I_y{}^i$ show up in $\mathcal{H}^{err}(t)$, and of course, also in $\mathcal{H}^{err}$. For $\alpha = \pm x$ (Gill–Meiboom modification) $\mathcal{H}^{err}$ commutes therefore with ρ_R (initial) and has no disturbing consequences. The opposite is true for $\alpha = y$ (original version of the CP train). For practical purposes this is probably the most important difference between the original and the Gill–Meiboom versions of the CP train. Obviously, $\mathcal{H}^{err}$ vanishes for both modifications (a) and (b).

$\mathcal{H}^{(1)}$ now contains cross terms between

$$\mathcal{H}^{inh}(t) = -\sum_i \Delta\omega^i [B(t) I_z{}^i + C(t) I_x{}^i]$$

(see Fig. 4-7) and $\mathcal{H}^{err}(t)$. Those involving $C(t)$ are smaller by roughly a factor t_W/t_c than those involving $B(t)$ unless the latter are zero by symmetry. The reason is simply that $C(t)$ is nonzero only during the pulses. We shall neglect the small type of cross terms henceforth.

In the original version of the CP train both $\mathcal{H}^{err}(t)$ and $B(t)$ are symmetric about the center of the cycle and produce, therefore, no cross terms of the large type—which is comforting.

Modification (a) does produce cross terms of the large type; they contain, however, only $I_x{}^i$ operators and do not trouble us for $\alpha = y$.

Modification (b) does not produce cross terms of the large type. This is a consequence of the following properties of the cycle:

(i)   symmetry of $\mathcal{H}^{err}(t)$ and $B(t)$ about the centers of the two phase-reversed subcycles of the full cycle; and

(ii) $\mathcal{H}^{inh} = 0$ in each subcycle.

In summary, the original version of the CP sequence is susceptible to finite pulse widths, flip angle errors, and rf inhomogeneity.

The Gill–Meiboom version and modification (a) are not susceptible to these imperfections in first order of t_W/t_c and ε, respectively; however, they are so only because the various correction terms commute with ρ_R (initial).

All the correction terms in question vanish for modification (b), which may be expected, therefore, to be superior to the other versions if only, perhaps, with regard to ease of adjustment and stability of the sequence.

2. FREQUENCY SHIFTS ARISING FROM cw rf FIELDS; BLOCH–SIEGERT SHIFT

It is well known that the presence of cw rf fields, off or on resonance with a spin system causes a "second-order" shift of the "resonance frequency." This effect can be studied in a straightforward manner by applying the average Hamiltonian theory. We put "resonance frequency" in parentheses to indicate that a certain problem is involved in its definition. The frequency shift arising from a linearly polarized on-resonance irradiation is called the Bloch–Siegert shift.[71] Off-resonance rf fields with substantial rf power are often applied in double-resonance experiments.

For the sake of clarity let us consider the following well-defined experimental situation. A (liquid) sample with spins I in a high magnetic field $\mathbf{B}_{st} = (0, 0, B_0)$ is irradiated continuously with a nonsaturating cw rf field $B_1 \cos(\omega t + \varphi)$ polarized linearly along the x-direction. At time $t = -t_W$ an intense P_x^{90} pulse of width t_W is applied such that it creates for $t > 0$ the initial condition $\rho(0) \propto I_y$. As before we partition B_0 into $B_0^I + \Delta B$. Because we want to cover also off-resonance cases we no longer equate $\gamma_n^I B_0^I$ with ω, but keep $\omega^I \equiv \gamma_n^I B_0^I$ (or $\Delta\omega = \gamma_n^I \Delta B$) as an adjustable parameter. The reason for introducing $\Delta\omega$ here is to have a means for monitoring the "resonance frequency" of the spins in the presence of the rf field. This somewhat cryptic statement will become clear presently. For $t > 0$ the Hamiltonian is

$$\mathscr{H} = \mathscr{H}_Z + \mathscr{H}_{\text{off}} + \mathscr{H}_{\text{rf}},$$

where

$$\mathscr{H}_Z = -\gamma_n^I B_0^I I_z = -\omega^I I_z,$$

$$\mathscr{H}_{\text{off}} = -\gamma_n^I \Delta B I_z = -\Delta\omega I_z,$$

$$\mathscr{H}_{\text{rf}} = -\gamma_n^I B_1 I_x \cos(\omega t + \varphi) = -2\omega_1 I_x \cos(\omega t + \varphi).$$

As before we set $\tfrac{1}{2}\gamma_n^I B_1 = \omega_1$. $U_Z = \exp[-i\mathscr{H}_Z t] = \exp[+i\omega^I I_z t]$ brings us into the rotating frame. The (untruncated) rotating frame Hamiltonian is

$$\mathscr{H}_R(t) = U_Z^{-1}(\mathscr{H}_{\text{off}} + \mathscr{H}_{\text{rf}}) U_Z$$

$$= -\Delta\omega I_z - 2\omega_1 \cos(\omega t + \varphi)[I_x \cos\omega^I t + I_y \sin\omega^I t]$$

$$= -\Delta\omega I_z - \omega_1 \{I_x[\cos((\omega^I + \omega)t + \varphi) + \cos((\omega^I - \omega)t - \varphi)]$$

$$+ I_y[\sin((\omega^I + \omega)t + \varphi) + \sin((\omega^I - \omega)t - \varphi)]\}. \tag{4-59}$$

We want to apply the average Hamiltonian theory to this time-dependent Hamiltonian. We immediately face a problem: $\mathscr{H}_R(t)$ is not necessarily periodic. It is so only if ω^I and ω are commensurable, that is, if $m/\omega^I = n/\omega$ for integer m and n. However, by keeping $\Delta\omega$ as a parameter we can always see that this condition is fulfilled. Recall that B_0 and ω are experimentally defined quantities, whereas ω^I and $\Delta\omega$ are artifacts of the theory. Hence we can always find a period t_c for $\mathscr{H}_R(t)$ given by $t_c = 2\pi n/\omega = 2\pi m/\omega^I$, where m and n are some integers.

[71] F. Bloch and A. Siegert, *Phys. Rev.* **57**, 522 (1940).

Obviously $\mathcal{H}_R$ differs for the off- and on-resonance cases. Therefore, we shall discuss them separately. We start with the on-resonance[72] case $\omega = -\omega^I$.

a. *On-Resonance Case*

In this case,

$$\mathcal{H}_R^{on}(t) = -\Delta\omega I_z - \omega_1 [I_x \cos\varphi + I_y \sin\varphi]$$
$$- \omega_1 [I_x \cos(2\omega^I t - \varphi) + I_y \sin(2\omega^I t - \varphi)]. \quad (4\text{-}60)$$

The period of $\mathcal{H}_R(t)$ is $t_c = \pi/\omega^I$, which is half the Larmor period of the I spins. The average or truncated Hamiltonian is given by the first line of Eq. (4-60). So far we have always restricted ourselves to this term.

Turning to $\overline{\mathcal{H}}^{(1)}$ we face the question of how to choose the cycle (its duration is fixed!). In view of the experimental difficulties related to such a choice, we avoid making any decision now and define the (first) cycle as running from τ to $\tau + \pi/\omega^I$, where τ is kept as a free parameter. Note that τ has a well-defined experimental meaning. A straightforward calculation yields

$$\overline{\mathcal{H}}_{on}^{(1)} = -\frac{i\omega^I}{2\pi} \int_\tau^{\tau + \pi/\omega^I} dt'' \int_\tau^{t''} dt' [\mathcal{H}_{R,on}(t''), \mathcal{H}_{R,on}(t')] \quad \text{(definition)}$$

$$= -\omega^I \left(\frac{\omega_1}{2\omega^I}\right)^2 [1 - 2\cos(2\omega^I\tau - 2\varphi)] I_z$$

$$- \Delta\omega \frac{\omega_1}{2\omega^I} \frac{1}{2} [I_x \cos(2\omega^I\tau - \varphi) + I_y \sin(2\omega^I\tau - \varphi)] \quad \text{(result)}.$$
$$(4\text{-}61)$$

$\overline{\mathcal{H}}_{on}^{(1)}$ is smaller than $\overline{\mathcal{H}}_{on}$ by roughly a factor ω_1/ω^I, so for many applications one is well justified to neglect $\overline{\mathcal{H}}_{on}^{(1)}$, and, of course, all further higher order correction terms. We note in passing that the term $-2\cos(2\omega^I\tau + 2\varphi)$ in the coefficient of I_z would be absent if only the "counterrotating" circularly polarized component of $\mathcal{H}_{rf}$ were present. On the other hand all correction terms $\overline{\mathcal{H}}^{(n)}$, $n > 0$, would vanish identically if only the "correct" component were present.

Now, what is the resonance frequency of the spins in an experiment as described above? In the absence of $\mathcal{H}_{rf}$, or when $\omega_1 \to 0$, the answer is clear: $\omega^I + \Delta\omega$. In other words, it is ω^I plus the coefficient of $-I_z$ in the expansion $\overline{\mathcal{H}}_{on} + \overline{\mathcal{H}}_{on}^{(1)} + \cdots$. By analogy we suspect that in the presence of $\mathcal{H}_{rf}$ the answer is

$$\omega_{res} = \omega^I + \Delta\omega + \omega^I(\omega_1/2\omega^I)^2 [1 - 2\cos(2\omega^I\tau - 2\varphi)]. \quad (4\text{-}62)$$

In a pulse experiment such as we are considering here, we tend to relate cumulative effects with ω_{res}. It is therefore disturbing that it should depend

---

[72] We need the minus sign to remain consistent with Section C,2,a. The final result, Eq. (4-64), does not depend on sign conventions.

on φ, and particularly on τ, which means the way we are defining the cycle. To overcome this difficulty let us go back one step and consider the motion of the magnetization $\langle \mathbf{M}(t) \rangle_R$ in the rotating frame after the pulse. To simplify matters we choose $\Delta\omega = -\omega^I(\omega_1/2\omega^I)^2$ and $\varphi = 0$. Note that an experimenter would be free to make such choices.

We are then left with

$$F = \bar{\mathscr{H}}_{\text{on}} + \bar{\mathscr{H}}_{\text{on}}^{(1)} + \cdots \approx -\omega_1 I_x + \omega_1 (\omega_1/2\omega^I) \cos 2\omega^I \tau I_z. \quad (4\text{-}63)$$

(We neglect terms of order $\omega_1 (\omega_1/\omega^I)^2$ as much smaller than those retained.) F is time independent but carries a periodic parametric dependence on τ. Note that it is the coefficient of I_z that carries this periodicity. Now think of the coefficients of I_x and I_z as representing magnetic fields in the rotating frame. Choosing the sampling points at

$$\tau = \frac{\pi}{4\omega^I} , \; ..., \; \frac{\pi}{4\omega^I} + N\frac{\pi}{\omega^I}$$

(i.e., such that $\cos 2\omega^I \tau = 0$) the effective field is along the x axis and, when observed stroboscopically, $\langle \mathbf{M}(t) \rangle_R$ will be seen to move in the yz plane (see Fig. 4-8).

Choosing the sampling points such that $\cos 2\omega^I \tau = \pm 1$, we find effective fields

$$\mathbf{B}_{\text{eff}} = \frac{1}{\gamma_n{}^I} \left(\omega_1, 0, \mp\omega_1 \frac{\omega_1}{2\omega^I} \right),$$

and $\langle \mathbf{M}(t) \rangle_R$ moves, when observed stroboscopically, in planes *tilted* somewhat with respect to the yz plane (see Fig. 4-8). It requires little imagination to guess what happens between: The magnetization moves in little wiggles with angular frequency $2\omega^I$. It must do so since it is under the influence of a

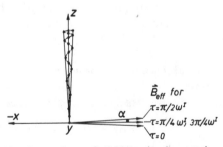

FIG. 4-8. Stroboscopic observations of $\langle \mathbf{M}(t) \rangle_R$ in the rotating frame. The angular frequency of the local oscillator for the phase-sensitive detector is ω^I. $\Delta\omega = \Delta B/\gamma_n{}^I$ is chosen as $-\omega^I(\omega_1/2\omega^I)^2$. The figure is drawn for $\alpha \approx \tan\alpha = \omega_1/2\omega^I = 0.05$. The frequency of the oscillations is $2\omega^I$ and increases only apparently toward z since the projection of the orbit $\langle \mathbf{M}(t) \rangle_R$ is shown.

driving force of angular frequency $2\omega^I$ [see Eq. (4-60)]. Note, however, that it would have been difficult to predict directly from Eq. (4-60) the direction in which the magnetization is going to oscillate.

On the average the magnetization remains in the yz plane and we conclude that the average resonance frequency $\bar{\omega}_{res}$ is all that can be defined reasonably and is given by

$$\bar{\omega}_{res} = \omega^I [1 + (\omega_1/2\omega^I)^2]. \tag{4-64}$$

The difference $\bar{\omega}_{res} - \omega^I = \omega^I (\omega_1/2\omega^I)^2$ is the well-known Bloch–Siegert shift.

This discussion makes clear that in the average Hamiltonian theory one must be really careful about defining cycles. This could be said to detract from the beauty of the theory. However, our "difficulty" actually arose from asking our question about ω_{res} at a too early stage. Had we inquired in the first place into $\rho_R(t)$ or into $\langle \mathbf{M}(t) \rangle_R$ we would not have run into any difficulty and the choices of φ and τ would have turned out to be essentially immaterial.

b. Off-Resonance Case[73]

Both $\mathscr{H}$ and $\mathscr{H}^{(1)}$ turn out to be even simpler than in the on-resonance case. Inspection of Eq. (4-59) and a straightforward calculation yield

$$\mathscr{H}_{off} = -\Delta\omega I_z, \tag{4-65}$$

$$\begin{aligned}
\mathscr{H}_{off}^{(1)} = &-I_z \frac{1}{2}\omega_1{}^2 \left\{ \frac{1}{\omega^I + \omega} + \frac{1}{\omega^I - \omega} \right\} \\
&- I_x \omega_1 \Delta\omega \left\{ \frac{\cos[(\omega^I + \omega)\tau + \varphi]}{\omega^I + \omega} + \frac{\cos[(\omega^I - \omega)\tau - \varphi]}{\omega^I - \omega} \right\} \\
&- I_y \omega_1 \Delta\omega \left\{ \frac{\sin[(\omega^I + \omega)\tau + \varphi]}{\omega^I + \omega} + \frac{\sin[(\omega^I - \omega)\tau - \varphi]}{\omega^I - \omega} \right\}.
\end{aligned} \tag{4-66}$$

The second-order frequency shift of the I-spin system is just the coefficient of $-I_z$ in $\mathscr{H}_{off}^{(1)}$ and therefore given by

$$\frac{1}{2}\omega_1{}^2 \left\{ \frac{1}{\omega^I + \omega} + \frac{1}{\omega^I - \omega} \right\} = \omega^I \frac{\omega_1{}^2}{(\omega^I)^2 - \omega^2}. \tag{4-67}$$

Note that this shift changes its sign depending on whether $|\omega^I|$ is larger or smaller than $|\omega|$. Note also that the Bloch–Siegert shift is obtained from the first term of the coefficient of $-I_z$ [i.e., from $\frac{1}{2}\omega_1{}^2/(\omega^I + \omega)$] in the special case $\omega^I = \omega$.

To get an idea of the size of the effect consider a proton–^{13}C decoupling experiment carried out in a field of 1.5 Tesla ($\cong 15$ kG) on a solid sample.

[73] A. Lösche, *Ann. Phys.* (*Leipzig*) [6] **20**, 178 (1957).

A typical value of B_1 is 6×10^{-3} Tesla ($\cong 60$ G). Under these conditions we must be aware of a second-order shift of about -0.27 ppm in the ^{13}C spectrum. Under most circumstances this little shift can be neglected. On the other hand, in a proton–^{19}F decoupling experiment with the same values of B_0 and B_1 the shift in the proton spectrum becomes as large as $+34.8$ ppm, which certainly cannot be neglected.

3. HETERONUCLEAR DECOUPLING[74]

One major purpose of heteronuclear decoupling is the removal of the effects of direct and indirect couplings between I and S spins from I-spin spectra. These effects tend to mask lineshifts arising from single I-spin interactions. For a long time, it has been well known that they can be suppressed by irradiating the sample with a linearly polarized rf field $B_1^S \cos \omega t$ at or close to the Larmor frequency of the S spins. In order to learn something about how well they can be suppressed we need to consider the Hamiltonian

$$\mathscr{H} = \mathscr{H}_Z^I + \mathscr{H}_Z^S + \mathscr{H}^{IS} + \mathscr{H}_{rf}^S,$$

where, in particular, $\mathscr{H}^{IS}$ represents the I–S spin–spin interactions, and

$$\mathscr{H}_{rf}^S = -\gamma_n^S B_1^S S_x \cos \omega t = -2\omega_1^S S_x \cos \omega t \qquad \text{(definition of } \omega_1^S).$$

The direct effect of this rf field on the I spins was considered in the previous subsection and is disregarded here. We partition, as before, the applied static field B_0 into $B_0^S + \Delta B$, where B_0^S is defined such that $\omega_S = \gamma_n^S B_0^S = -\omega$. The idea behind this partitioning is the intention to study heteronuclear decoupling with a monochromatic rf field not exactly on resonance with the S spins.

Accordingly, we write

$$\mathscr{H}_Z^S = -\omega_S S_z - \Delta\omega S_z \qquad (\Delta\omega = \gamma_n^S \Delta B \ll \omega_S).$$

There is no need to partition $\mathscr{H}_Z^I$ in a similar manner. By way of the interaction representation $U_Z = \exp[-i(\mathscr{H}_Z^I - \omega_S S_z)t]$ we switch over to the rotating I- and S-spin frames. Note that these frames rotate with different angular velocities about the laboratory frame z axis. The truncated rotating frame Hamiltonian is

$$\mathscr{H}_R = \mathscr{H}_{off} + \mathscr{H}_{rf,R}^S + \mathscr{H}_R^{IS},$$

[74] In high-resolution NMR in liquids one contrasts heteronuclear with homonuclear decoupling. The difference is, however, only quantitative or at most technical. What counts is that the observed (I) and perturbing (S) spins have different NMR frequencies, either because they belong to different isotopes (heteronuclear decoupling) or because they are differently chemically shifted (homonuclear decoupling). By contrast, the observed and perturbing spins are identical in WAHUHA experiments.

where

$$\mathcal{H}_{\text{off}} = -\Delta\omega S_z, \qquad \mathcal{H}_{\text{rf, R}}^S = -\omega_1{}^S S_x, \qquad \mathcal{H}_R^{IS} = \sum_{i,k} A^{ik} I_z{}^i S_z{}^i,$$

with

$$A^{ik} = J^{ik} + J_{zz}^{(2)i,k} - \gamma_n{}^I \gamma_n{}^S \hbar D_{zz}^{ik}.$$

In liquids we are primarily concerned with J^{ik}, whereas in solids we have to fight D_{zz}^{ik}, which is proportional to r_{ik}^{-3}.

$\mathcal{H}_R^{IS}$ transforms in both the I- and S-spin spaces as a first-rank tensor (cf. Table 4-1). Hence a rotation in S-spin space about the first-rank magic angle $\beta_m^{(1)} = 90°$ will result in a zero average of the I–S-coupling term. The required motion is imposed on the S-spin operators by

$$U_{\text{rf}} = \exp[i\omega_1{}^S S_x t].$$

The combination of U_Z and U_{rf} defines a rotating I-, doubly rotating S-spin frame of reference. The effective Hamiltonian in this frame is

$$\tilde{\mathcal{H}}(t) = U_{\text{rf}}^{-1}(\mathcal{H}_R^{IS} + \mathcal{H}_{\text{off}}) U_{\text{rf}} \qquad \text{(definition)}$$

$$= \sum_{i,k} A^{ik} I_z{}^i (S_z{}^k \cos\omega_1{}^S t - S_y{}^k \sin\omega_1{}^S t)$$

$$- \Delta\omega(S_z \cos\omega_1{}^S t - S_y \sin\omega_1{}^S t). \tag{4-68}$$

$\tilde{\mathcal{H}}(t)$ is periodic, with period $t_c = 2\pi/\omega_1{}^S$. The average of $\tilde{\mathcal{H}}(t)$ over t_c vanishes as we have anticipated. This holds true regardless of the size of $\omega_1{}^S$ or $B_1{}^S$; however, it does not yet tell us much about the efficiency of the decoupling scheme. Therefore, we must inquire into $\bar{\mathcal{H}}^{(1)}$ and, eventually, also into $\bar{\mathcal{H}}^{(2)}$. A simple straightforward calculation yields

$$\bar{\mathcal{H}}^{(1)} = -\frac{i}{2t_c} \int_0^{t_c} dt'' \int_0^{t''} dt' [\tilde{\mathcal{H}}(t''), \tilde{\mathcal{H}}(t')] \qquad \text{(definition)}$$

$$= -\frac{1}{2\omega_1{}^S} \sum_{i,m,k} A^{ik} A^{mk} I_z{}^i I_z{}^m S_x{}^k + \frac{\Delta\omega}{\omega_1{}^S} \sum_{i,k} A^{ik} I_z{}^i S_x{}^k - \frac{\Delta\omega^2}{2\omega_1{}^S} \sum_k S_x{}^k. \tag{4-69}$$

The last term is the simplest: It does not affect the I-spin system. The middle one vanishes when the S spins are irradiated exactly on resonance ($\Delta\omega = 0$). We postpone for a while the discussion of this term for $\Delta\omega \neq 0$ and turn immediately to the first term, which deserves a little more attention: For fixed i the summation is over all other spins m of the I-spin system and over all spins k of the S-spin system. A special situation arises when the spin quantum number of the I spins is $\frac{1}{2}$. This is so in virtually all cases of practical interest (^{13}C, ^{15}N, ^{1}H, ^{19}F, etc.). For $I = \frac{1}{2}$ and $i = m$ we have $I_z{}^i I_z{}^m = (I_z{}^i)^2 = \frac{1}{4}$, independent of the representation. This means that the $i = m$ term in the sum $\sum_{i,m,k}$ has no influence at all on the I-spin spectrum. Next consider the terms $i \neq m$. To be specific let us have in mind a proton-decoupled ^{13}C-experiment on a solid sample in which hydrogen atoms are bonded directly to the carbons of interest. Select a certain ^{13}C spin as spin i. The dominant part of A^{ik} is proportional to r_{ik}^{-3} and therefore (relatively) large only if k represents a directly bonded proton. A^{mk} is proportional to r_{mk}^{-3}, where r_{mk} is the distance of the kth proton to another ^{13}C nucleus. In a sample with ^{13}C in natural abundance r_{mk} will typically be much larger than the C—H bond distance. As a result A^{mk} will be very small and we may expect that the entire first term of $\mathscr{H}^{(1)}$ affects the ^{13}C spectrum—in particular, the widths of the resonance lines—to a negligible extent only. The situation changes somewhat in samples highly enriched with ^{13}C. The change, however, is not very dramatic since the second smallest proton–carbon distance in, e.g., benzene is still more than twice as large as the smallest one.

In short, $\mathscr{H}^{(1)}$ is probably never the resolution-limiting term in proton-decoupled ^{13}C spectroscopy on solid samples and the same holds true, a fortiori, for proton-decoupled ^{15}N spectroscopy.

To find the resolution-limiting term we must go one step further in the Magnus expansion. However, in calculating the next term, $\mathscr{H}^{(2)}$, we can restrict the sum over i, k in $\mathscr{H}(t)$ [cf. Eq. (4-68)] to the one single term that describes the spin–spin coupling of an ^{1}H—^{13}C or ^{1}H—^{15}N fragment. The necessary integrations are straightforward (not even very lengthy) and the result is

$$\mathscr{H}^{(2)} = \frac{1}{6}\frac{1}{(\omega_1{}^S)^2}A^3 I_z{}^3 S_z = \frac{1}{24}\frac{A^2}{(\omega_1{}^S)^2}A I_z S_z, \qquad (4\text{-}70)$$

where A is the coupling constant of the ^{1}H—^{13}C or ^{1}H—^{15}N fragment.

This result lends itself to a very simple interpretation. Provided $A/\omega_1{}^S < 1$ the leading proton–carbon interaction term in the rotating frame ($A I_z S_z$) is reduced by the irradiating rf field at resonance with the proton spins by the factor $\frac{1}{24}(A/\omega_1{}^S)^2$.

The members of the MIT NMR group have indeed observed in their pioneering ^{13}C–proton decoupling experiments on solid samples[50] that $\omega_1{}^S$ need not surpass very much the strongest coupling coefficient $A^{ik} = A$ for obtaining good results, which means very narrow ^{13}C-resonance lines. Furthermore, they noticed that the crossover from poor to acceptable decoupling is fairly sharp. Our Eq. (4-70) predicts it to be quadratic in $A/\omega_1{}^S$.

$\Delta\omega \neq 0$; Noise Decoupling; Chirping Pulses

It is clear from the second term in Eq. (4-69) that on-resonance irradiation of the S spins assures the maximum decoupling efficiency. Nuclear magnetic shielding and quadrupole interactions of the S spins often cause, however, a considerable spread of the S-spin resonances. It is then impossible with a monochromatic decoupling rf field to be on resonance with all the S spins. Then one must either live (or die!) with the $(\Delta\omega/\omega_1{}^S)\,A^{ik}I_z{}^iS_x{}^k$ term, trying to make it as small as possible by making $\omega_1{}^S$ as large as possible, or one must resort to one of the more sophisticated decoupling procedures to be briefly mentioned now.

Noise decoupling approaches the problem by a random (often pseudo-random, i.e., in the long run repetitive) phase, amplitude, or frequency modulation of the rf field. The idea is to generate a band of decoupling frequencies and thereby to provide for each individual S spin an rf field at the most efficient decoupling frequency.

The same goal can be reached at by a string of "chirping" pulses. A chirping pulse is one in which the carrier frequency is swept through a certain (adjustable) range of frequencies from the beginning to the end of the pulse. The attractive feature of chirping pulses is that the intensity distribution of the generated rf spectrum can be tailored—at least in principle—according to the special needs of the sample at hand.

Noise decoupling is currently used extensively, particularly in ^{13}C NMR on both liquid and solid samples. The exploration of chirping pulses, on the other hand, is still in its planning phase. It should be pointed out, however, that neither of these schemes seems to be capable at the present time of decoupling in *solid* samples spin-$\frac{1}{2}$ nuclei, in particular protons and ^{13}C, from quadrupolar nuclei such as ^{2}H or ^{14}N whose spectra are spread over typical ranges of 170–200 kHz, and 4–5 MHz, respectively.

Deuteron decoupling has been achieved in nematic solutions by exploiting *double-quantum* transitions.[75a] It remains to be seen whether these techniques can also be applied to solids.[75b]

4. Heteronuclear Decoupling while Performing a WAHUHA Experiment

We have stated above that the truncated I–S coupling Hamiltonian in the rotating frame behaves as a first-rank tensor in I-spin subspace. Its average, therefore, does not vanish in a pure WAHUHA experiment (see Table 4-1). The I–S coupling is reduced by a factor $1/\sqrt{3}$; the quantity in which we are usually interested—the chemical shift—is, however, reduced by the same factor so we really gain nothing. A means of removing this term from the effective Hamiltonian is irradiating the S spins with an appropriate rf field at the Larmor frequency of the S spins. Figure 4-9 shows a number of possibilities, one of which is suggestive (A) but does not work, whereas the other two (B, C) do work.

[75a] R. C. Hewitt, S. Meiboom, and L. C. Snyder, *J. Chem. Phys.* **58**, 5089 (1973).
[75b] *Note added in proof:* Since the completion of this manuscript they have been applied successfully to solids, e.g., to ice (A. Pines, private communication).

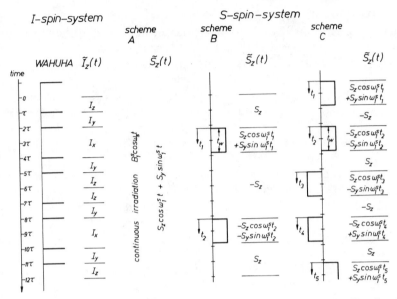

FIG. 4-9. Schemes for eliminating the I–S coupling term from the effective Hamiltonian while performing an I-spin WAHUHA experiment. $\omega_1{}^S t_w = \pi$ for schemes B and C. The average of $\tilde{I}_z(t)\,\tilde{S}_z(t)$ over two WAHUHA cycles vanishes for schemes B and C, but does not vanish, in general, for scheme A. Note that virtually the entire large window of the WAHUHA train is available for the S spin–π pulses.

Continuous irradiation of the S spins (scheme A) does not do the job unless

$$2\pi/\omega_1 = 2\tau/n, \quad n \text{ integer.}$$

If this condition is not met, the average of $\tilde{I}_z(t)\,\tilde{S}_z(t)$ does not vanish over the WAHUHA cycle and, even worse, this average varies from "cycle" to "cycle." We put cycle in quotes because under these conditions the WAHUHA cycle is no longer a true cycle.

A sequence of S spin–180° pulses in either type of large windows of the WAHUHA sequence does the job (scheme B) even when the width of the S spin–180° pulses is finite. One glance at Fig. 4-9 will convince the reader of this fact. Note, however, that the cycle time is 12τ instead of 6τ. The complementary type of large windows can be chosen as convenient observation windows.

There may be cases where one would like to pulse the S spins at a higher rate in order to obtain a better averaging efficiency. It is not possible simply to exploit both types of large WAHUHA windows for S spin–180° pulses. The average of the coefficients of $I_z S_y$ and $I_x S_y$ would not vanish. Reversing the rf phase of alternate pairs of S spin–180° pulses cures the trouble (scheme C).

The superiority of scheme C over scheme B becomes fully clear if we disregard the $S_y \sin \omega_1{}^S t_n$ parts of $\tilde{S}_z(t)$. This we can do for $t_W/\tau \ll 1$. For scheme B, $\tilde{\mathscr{H}}_R^{IS}(t) \propto \tilde{I}_z(t)\tilde{S}_z(t)$ becomes symmetric around the center 6τ of the cycle with the effect that in addition to $\langle \tilde{\mathscr{H}}_R^{IS}(t)\rangle_{av} = 0$ all correction terms of odd order vanish. For scheme C, on the other hand, $\tilde{\mathscr{H}}_R^{IS}(t)$ becomes antisymmetric and $\langle \tilde{\mathscr{H}}_R^{IS}(t)\rangle_{av}$ vanishes together with all correction terms, odd and even (see Chapter IV, Section D,4).

Mehring et al.[76] have performed a first successful homoheteronuclear decoupling experiment on NaF. Scheme B was used. The ^{19}F spins were observed, with the ^{23}Na spins playing the role of the S spins. A ^{19}F NMR line with a width of about 4 ppm, shifted by $+57$ ppm with respect to C_6F_6, was obtained. The WAHUHA sequence alone failed to produce a sizably narrowed ^{19}F spectrum. At the end of the previous subsection we pointed out that decoupling of any kind of spins from quadrupolar nuclei such as ^{23}Na is a difficult if not insurmountable problem. In NaF, however, the Na ions sit on cubic lattice sites where the field gradient vanishes. As a result, there is no quadrupolar spread of ^{23}Na-NMR frequencies in NaF. This circumstance was crucial for the success of the experiment of Mehring et al.

Further homoheteronuclear decoupling experiments have been reported by Mehring et al.[77] (I spins: ^{19}F, S spins: ^{1}H), Burghoff et al.[78] (I spins: ^{1}H, S spins: ^{31}P), and Van Hecke et al. (I spins: ^{1}H, S spins: ^{19}F).[79]

5. SELF-DECOUPLING

Recall how heteronuclear decoupling works: By external means we introduce a term into the spin Hamiltonian, which affects the S spins only. What results is a time dependence of the S spin operators over which we may average eventually. Now, there are automatically terms in the total nuclear spin Hamiltonian that contain S spin operators only. Two important examples are $\mathscr{H}_D^{SS}$ and $\mathscr{H}_Q^{S}$, which describe, respectively, the omnipresent direct spin–spin couplings among the S spins and the quadrupolar couplings of the S spins with electric field gradients, provided $S \geqslant 1$. When molecular motions are present in the sample both $\mathscr{H}_D^{SS}$ and $\mathscr{H}_Q^{S}$ depend explicitly on the time t.

With regard to an I-spin experiment we may treat $\mathscr{H}_D^{SS}$ and $\mathscr{H}_Q^{S}$ in an analogous manner as we treated $\mathscr{H}_{rf}^{S}$, that is, we may remove these terms

---

[76] M. Mehring, A. Pines, W. -K. Rhim, and J. S. Waugh, J. Chem. Phys. 54, 3239 (1971).

[77] M. Mehring, H. Raber, and G. Sinning, Proc. Congr. AMPERE, 18th, 1974 1, 35 (1974).

[78] U. Burghoff, G. Scheler, and R. Müller, Phys. Status Solidi. A 25, K31 (1974); U. Burghoff, H. Rosenberger, R. Zeiss, R. Müller, and L. N. Rashkovich, ibid. 26, K171 (1974).

[79] P. Van Hecke, H. W. Spiess, and U. Haeberlen, J. Magn. Resonance, in press.

from the Hamiltonian by way of the interaction representation

$$U_S(t) = T \exp\left\{-i \int_0^t [\mathscr{H}_D^{SS}(t') + \mathscr{H}_Q^S(t')]\, dt'\right\}. \tag{4-71}$$

A consequence of $U_S(t)$ is a time dependence of $\mathscr{H}_R^{IS}$, or of S_z:

$$S_z \to \tilde{S}_z(t) = U_S^{-1}(t)\, S_z\, U_S(t). \tag{4-72}$$

The time dependence of $\tilde{S}_z(t)$ is random and therefore characterized by a correlation time τ_c rather than by a period. This constitutes an important new feature that complicates the quantitative aspects of self-decoupling. Qualitatively, however, it goes without saying that $\tilde{S}_z(t)$ can eventually be replaced by an average. We expect that "self-decoupling" becomes effective when τ_c^{-1} becomes larger than the "strength" of the I–S coupling.

In order to get a feeling for the nature and size of τ_c and for the practical importance of self-decoupling we consider now a few examples.

a. Free Induction Decay and Standard cw Experiments

Abragam[9] (Chapter IV) describes experiments on solid KF (I spins: ^{39}K, S spins: ^{19}F) and on solid AgF (I spins: ^{109}Ag, S spins: ^{19}F), where self-decoupling due to $\mathscr{H}_D^{FF}$ is strong enough to reduce the I-spin linewidth by as much as a factor of ten. He states that at least the ^{109}Ag resonance could not have been observed if self-decoupling had not been operative. Abragam, however, uses a different approach to describe the effect.

Our next example is proton self-decoupling in ^{13}C and ^{15}N NMR in solids. Only $\mathscr{H}_D^{HH}$ is operative. An estimate for τ_c^{-1} can be made on the basis of the second moment $\langle \Delta\omega^2 \rangle$ of the proton resonance line. According to Eq. (3) of Abragam[9], Chapter V,

$$\tau_c^{-1} = W = \langle \Delta\omega^2 \rangle^{1/2}/30. \tag{4-73}$$

In a typical organic solid $\langle \Delta\omega^2 \rangle$ is about 1.4×10^{10} sec^{-2} ($\hat{=} 20$ G^2). A typical number for τ_c^{-1} is therefore 4×10^3 sec^{-1}.

A measure for the proton–carbon coupling "strength" is the coefficient of $I_z S_z$ in $\mathscr{H}_R^{IS}$, which we divide by 2π in order to get a quantity suitable for comparison with τ_c^{-1}:

proton–carbon coupling strength

$$= \frac{1}{2\pi} \hbar \gamma_n^{\,H} \gamma_n^{\,C} r^{-3} \left(\frac{4}{5}\right)^{1/2} = 2.7 r^{-3} \times 10^4 \text{ sec}^{-1},$$

where r is the proton–carbon interatomic distance in angstroms. The corresponding number for ^{15}N is $1.1 r^{-3} \times 10^4$ sec^{-1}. We have taken a powder average over the orientation dependence of $\mathscr{H}_R^{IS}$, which means we replaced

$1 - 3 \cos^2 \Theta$ by

$$\langle (1 - 3 \cos^2 \Theta)^2 \rangle_{av}^{1/2} = \sqrt{\tfrac{4}{5}}.$$

The same average is incorporated in our estimate of τ_c^{-1}. We conclude that we may expect a substantial narrowing effect if the distance between the carbon (or nitrogen) and the nearest proton exceeds, say, 2.7 Å. Such minimum distances are not uncommon in nature.

A practical example where self-decoupling of this type appears to be operative is triethanolamine, $(OHCH_2CH_2)_3{}^{15}N$. At liquid nitrogen temperature we found $\langle \Delta\omega^2({}^{15}N) \rangle = 3 \times 10^7$ sec^{-2}, whereas the theoretical value calculated on the basis of Eq. (IV-55) in Abragam[9] is 10^8 sec^{-2}. The discrepancy may even be larger than expressed by these two numbers since the experimental value contains contributions of unknown size from ^{15}N nuclear magnetic shielding anisotropy. We believe that this is a manifestation of proton self-decoupling. As a result of an unusually high proton packing density the second moment of the proton resonance is particularly large in triethanolamine; our experimental value is 2.65×10^{10} sec^{-2} ($\hat{=} 37$ G^2).

b. Multiple-Pulse Experiments

The realm of multiple-pulse experiments is ^1H and ^{19}F NMR in solids (see Chapter VI). If the S spins are either ^{19}F or ^1H, the coupling strength will typically be so large that no self-decoupling occurs. If the S spins are, on the other hand, other spin-$\tfrac{1}{2}$ spins, $\mathcal{H}_D^{SS}$ will be typically so small that W becomes too small for self-decoupling. Therefore, we may expect that $\mathcal{H}_D^{SS}$ is seldom if ever an efficient source for self-decoupling. What remains as a possibility is self-decoupling driven by quadrupolar couplings $\mathcal{H}_Q{}^S$ of spins with $S \geqslant 1$. Here, the correlation time τ_c becomes equal to the spin lattice relaxation time of the S spins. (We assume that a uniform spin lattice relaxation time of the S spins exists.)

Since active heteronuclear decoupling of quadrupolar nuclei in noncubic environments seems unfeasible at the moment, self-decoupling of this type is very attractive. An example where it is operative is trans-diiodoethylene, $C_2H_2I_2$. A WAHUHA multiple-pulse experiment that we carried out at 90 MHz yielded a powder spectrum with a width of not more than 400 Hz. The entire spread of the olefinic proton shift anisotropy is contained in that number. Without iodine self-decoupling the WAHUHA proton spectrum would have had a width in excess of 2.4 kHz. Further measurements on single crystals[80] led to substantially narrower lines but revealed, in addition, a critical dependence of the self-decoupling efficiency on the coupling strength and on W. In single crystals the coupling strength does vary as $1 - 3 \cos^2 \Theta$, and W is also dependent, in general, on the orientation of the crystal.

[80] See also Chapter VI, Section C, 2 and Fig. 6-9.

V. Detailed Discussion of Multiple-Pulse Sequences Intended for High-Resolution NMR in Solids

With extremely few exceptions, researchers in high-resolution NMR in liquids at present buy their spectrometers from commercial manufacturers. One of the questions a spectrometer salesman will invariably be asked by a potential customer is: What resolution can your instrument achieve? The answer the customer expects—and invariably gets without hesitation—is a number below 0.3 Hz.

Probably every experimenter who has made an effort in high-resolution NMR in solids by either multiple-pulse or magic-angle sample-spinning techniques has been confronted with the same question on repeated occasions. Probably on no such occasion has he answered without hesitation. The reason is twofold (at least). First, he knows that his answer will be compared with something like 0.3 Hz, and that the number he will quote eventually will be much larger than that. Second, he feels compelled to explain that and why his answer cannot be a general one. It depends largely on the kind, size, shape, and orientation of the sample at hand. If he works with multiple-pulse sequences, he finally will give a number for a single crystal of CaF_2 oriented with its 111 direction parallel to the applied field, and the number will be somewhere in the range of 15 to 100 Hz.

What we intend to illustrate with these remarks is

(1) Resolution is a problem in high-resolution NMR in solids and can be expected to remain one for the foreseeable future.

(2) In contrast to high-resolution NMR in liquids, where resolution is just

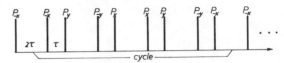

FIG. 5-1. MREV eight-pulse cycle.

a matter of the homogeneity of the applied magnetic field,[81] the problem is very complex in high-resolution NMR in solids.

(3) The main efforts of advancing high-resolution NMR in solids are still done in research laboratories.

There is a large and still growing variety of schemes designed for achieving high-resolution NMR in solids. However, only two seem to be currently in use on a routine basis for actual data collection. These are the WAHUHA four-pulse cycle, which has already served us for introducing the subject, and an eight-pulse cycle first proposed[82] by Mansfield[83] and first applied successfully to CaF_2 by Rhim et al.[67] This sequence is depicted in Fig. 5-1. We shall discuss these two sequences in some detail. In doing so we shall encounter a variety of aspects that bear on the practical usefulness of a line-narrowing multiple-pulse sequence, e.g., resolution, scaling factor, signal to noise ratio, sensitivity to misalignments of pulses, and observability of the NMR signal. In Section F we shall briefly review further propositions for high-resolution NMR in solids.

For what follows it is useful to define ideal pulse sequences, which consist of pulses with width $t_W \to 0$. Furthermore, the pulses are assumed to be perfect with respect to rf homogeneity, nutation angle, nutation axis, spacing, etc. This enumeration may give already some impression of what kind of problems we have to deal with in practice.

A. Properties of the Ideal WAHUHA Four- and MREV Eight-Pulse Cycles

In Chapter IV, Section C,2, we discussed the lowest order or average Hamiltonian approximation of the WAHUHA sequence. We stated that it is adequate for $t_c \| \mathcal{H}_{sec}^{int} \| \ll 1$. This condition, however, tells us nothing about

---

[81] To be sure, we admire greatly the achievements of the NMR industry, which is now able to offer magnetic fields homogeneous to within 1 part in 10^9 over sample volumes as large as about 1 cm³!

[82] Mansfield described this cycle in a shorthand but somewhat esoteric notation. This may be the reason why his fatherhood of the cycle remained unnoticed for some time. By the way, we shall not term pulse sequences with different preparation pulses as different.

[83] P. Mansfield, J. Phys. C 4, 1444 (1971).

the line-narrowing efficiency of the cycle. Therefore, we must study the correction terms as they follow from the average Hamiltonian theory. In this section we shall evaluate the first- and second-order correction terms for the ideal WAHUHA and MREV sequences, but we shall also summarize those properties of these cycles that follow directly from the average Hamiltonian.

1. WAHUHA FOUR-PULSE CYCLE

a. *Properties Related to the Detection of the NMR Signal*

The initial amplitude of the oscillating part of the NMR signal is two-thirds the initial amplitude of a free-induction decay signal unless a special 45° preparation pulse is used. This means for typical applications of the WAHUHA sequence that the signal-to-noise ratio is substantially smaller than it could be.

There is a convenient 2τ window for sampling the NMR signal.

b. *Properties of Zeroth-Order Average Hamiltonian*

Scalar (zeroth-rank) I-spin–I-spin interactions remain unaffected. First-rank I-spin interactions are retained, but scaled down by a factor of $1/\sqrt{3}$.

The average of second-rank I-spin–I-spin interactions vanishes. The smallest interval of time for which this average vanishes is $3\tau = t_c/2$.

c. *First-Order Corrections*

The cycle is symmetric (see Fig. 4-2); hence $\mathscr{H}^{(n)} = 0$ for n odd, and in particular $\mathscr{H}^{(1)} = 0$. This means the Magnus expansion

$$F = \bar{\mathscr{H}} + \bar{\mathscr{H}}^{(1)} + \bar{\mathscr{H}}^{(2)} + \cdots$$

contains no purely dipolar terms quadratic in $\mathscr{H}_D$, nor cross terms of the type $\mathscr{H}_D \mathscr{H}_{CS}$ or $\mathscr{H}_D \mathscr{H}_{off}$.

d. *Second-Order Corrections*

The evaluation of the second-order corrections $\bar{\mathscr{H}}^{(2)}$ to the average Hamiltonian $\bar{\mathscr{H}}$ seems to be a formidable task in view of the threefold integrals involved. In fact, it is not, at least not for ideal pulse cycles for which the threefold integrals reduce to simple sums. For symmetric cycles such as the WAHUHA cycle, further simplifications arise (see Fig. 5-2): From the symmetry of $\bar{\mathscr{H}}^{(2)}$ with respect to $\mathscr{H}(t_3)$ and $\mathscr{H}(t_1)$—they can be interchanged; see Eq. (4-41)—it follows that $I_{\text{①}} = I_{\text{③}}$ and $I_{\text{②}} = I_{\text{④}}$, where $I_{\text{⊗}}$ is the contribution of domain ⊗ to $\bar{\mathscr{H}}^{(2)}$. This is indicated in Fig. 5-2. Hence it is sufficient to restrict the integration over domains ① and ②. $I_{\text{②}}$ (and $I_{\text{④}}$) vanish, moreover, when in addition to

$$\mathscr{H}(t) = \mathscr{H}(t_c - t) \qquad \text{(condition a: symmetry of the cycle),}$$

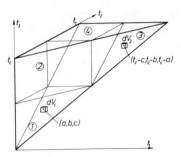

FIG. 5-2. Second-order average Hamiltonian for symmetric cycles. The symmetry of the cycle and the invariance of

$$I = \overline{\mathscr{H}}^{(2)} = \frac{-1}{6t_c} \int_0^{t_c} dt_3 \int_0^{t_3} dt_2 \int_0^{t_2} dt_1 \{[\mathscr{H}(t_3), [\mathscr{H}(t_2), \mathscr{H}(t_1)]] + [\mathscr{H}(t_1), [\mathscr{H}(t_2), \mathscr{H}(t_3)]]\}$$

against a permutation of $\mathscr{H}(t_1)$ and $\mathscr{H}(t_3)$ imply that the integrand of I is identical in volume elements dV_1 and dV_2 located at $(t_1, t_2, t_3) = (a, b, c)$ and $(t_c - c, t_c - b, t_c - a)$. If (a, b, c) is an arbitrary point in domain ① (domain ②), $(t_c - c, t_c - b, t_c - a)$ is in domain ③ (domain ④). It follows that $I_1 = I_3$ and $I_2 = I_4$.

we have

$$\overline{\mathscr{H}} = 0 \qquad \text{(condition b: zero-average Hamiltonian)}.$$

Proof. For domain ②, t_3 is not an integration limit of the integrals over t_1 and t_2. Hence, the integration over t_3 can be carried out separately. Conditions (a) and (b) imply that $\int_{t_c/2}^{t_c} \mathscr{H}(t_3) \, dt_3 = 0$. Both conditions (a) and (b) are met for the purely dipolar second-order correction term, given by

$$\overline{\mathscr{H}}_D^{(2)} = \frac{1}{648} t_c^2 [(\mathscr{H}_D^x - \mathscr{H}_D^z), [\mathscr{H}_D^x, \mathscr{H}_D^y]], \qquad (5\text{-}1)$$

where

$$\mathscr{H}_D^z = \gamma_n^2 \hbar \sum_{i<k} \frac{1}{2} (3 \cos^2 \vartheta_{ik} - 1)(1/r_{ik}^3)(\mathbf{I}^i \cdot \mathbf{I}^k - 3I_z^i I_z^k).$$

$\mathscr{H}_D^x$ and $\mathscr{H}_D^y$ are expressions analogous to $\mathscr{H}_D^z$ with the indices z replaced by x and y, respectively (see Table 4-2).

The damping constant associated with $\overline{\mathscr{H}}_D^{(2)}$ is proportional to $T_{2,\text{rigid}}(T_{2,\text{rigid}}/t_c)^2$, where $T_{2,\text{rigid}}$ is defined operationally, as before, by $\langle \Delta\omega^2 \rangle^{-1/2}$. The coefficient of this lowest order nonvanishing correction term,

$$\frac{1}{648} = \frac{1}{18} \left(\frac{\tau}{t_c}\right)^2,$$

tells us that the suppression of dipolar interactions starts to become effective as soon as τ—not t_c—becomes shorter than $T_{2,\text{rigid}}$. In view of the results of the early solid-echo experiments[84-87] this is but natural—in retrospect.

An interesting point to note is that $\mathscr{H}_D^{(2)}$ vanishes for so-called two-spin systems.[88] Otherwise stated, $\mathscr{H}_D^{(2)}$ contains no terms proportional to $(1/r_{ik}^3)^3$ but only terms proportional to $(1/r_{ik}^3)^2(1/r_{il}^3)$, $l \neq k$, etc. In substances with pairs of closely neighbored protons, e.g., methylene groups or water molecules, the pair interatomic distance r_{ik} is considerably smaller than all other interatomic distances. The absence of terms $(1/r_{ik}^3)^3$ in $\mathscr{H}_D^{(2)}$ means that the potentially largest terms, as far as geometry is concerned, actually do not contribute to $\mathscr{H}_D^{(2)}$ and, hence, to the broadening of the spectral lines.

The offset second-order correction term is

$$\mathscr{H}_{\text{off}}^{(2)} = -\Delta\omega(\Delta\omega^2\tau^2/18)\{(I_x+I_y+I_z) + 3(I_z-I_x)\}. \qquad (5\text{-}2)$$

The rotation imposed on the spins by the $(I_x+I_y+I_z)$ part is along the same axis as results from the zeroth-order term. It therefore modifies—slightly—the scaling factor. Instead of $1/\sqrt{3}$ it now becomes $(1/\sqrt{3})(1 - \Delta\omega^2\tau^2/6)$.

As the sampling rate in a WAHUHA experiment is $1/t_c$ the maximum frequency (Nyquist frequency ν_N) in the spectrum is $\nu_N = 1/2t_c = 1/12\tau$. The frequency-dependent correction to the scaling factor therefore never exceeds $4\pi^2 \cdot 3/6 \cdot 12^2 = 0.138 \cong 13.8\%$. (The factor 3 enters because the spectral frequencies are scaled frequencies, whereas $\Delta\omega$ is the unscaled off-resonance frequency.) Our experiments are usually arranged so that the highest frequency of interest in the spectrum is below $\frac{1}{2}\nu_N$, where the correction is about 3.5%. In our work on protons we even remain below $\frac{1}{4}\nu_N$—there the correction is negligible.

The rotation of the spins due to the (I_z-I_x) part in Eq. (5-2) is perpendicular to the main rotation. It therefore affects the scaling factor only in second order. However, it also tilts the rotation axis away from the rotating frame 111 axis. The maximum tilt angle is 18.5°, which applies for the Nyquist frequency.

It is a matter of course that everything we have said about the offset second-order correction term applies equally well to the chemical shift second-order correction term.

Cross terms. It is clear that there is a multitude of cross terms involving $\mathscr{H}_{\text{CS}}$, $\mathscr{H}_{\text{off}}$, $\mathscr{H}_D$, and $\mathscr{H}_J$. The cross term quadratic in $\mathscr{H}_D$ and linear in $\mathscr{H}_{\text{off}}$—

[84] I. G. Powles and P. Mansfield, *Phys. Lett.* **2**, 58 (1962).

[85] I. G. Powles and I. H. Strange, *Proc. Phys. Soc., London* **82**, 6 (1963).

[86] R. Hausser and G. Siegle, *Phys. Lett,* **19**, 356 (1965).

[87] G. Siegle, *Z. Naturforsch. A* **21**, 1722 (1966).

[88] B. Bowman, M.S. Thesis, Massachusetts Institute of Technology, Cambridge, 1969 (unpulished).

presumedly the most important one—is given by

$$(\tau^2 \, \Delta\omega/18)\{2[\mathscr{H}_D{}^x, [\mathscr{H}_D{}^y, I_x]] + 2[\mathscr{H}_D{}^z, [\mathscr{H}_D{}^x, I_y]]$$
$$- [\mathscr{H}_D{}^x, [\mathscr{H}_D{}^x, I_y]] + [I_z, [\mathscr{H}_D{}^x, \mathscr{H}_D{}^y]]\}. \tag{5-3}$$

The damping constant associated with this term is proportional to

$$T_{2,\,\text{rigid}}\left(\frac{T_{2,\,\text{rigid}}}{\tau} \cdot \frac{1}{\tau \, \Delta\omega}\right).$$

Note that the damping increases linearly with the offset! We avoid writing expressions for other cross terms because they can be expected not to limit the resolution in actual experiments.

2. MREV EIGHT-PULSE CYCLE

This cycle was first described by Mansfield[83] as one of an entire family of eight-pulse, 12τ complementary cycles that compensate for both finite pulse width and rf inhomogeneity effects. Rhim et al.[67] first used it in 1973 in actual experiments and got amazingly good line-narrowing results. Later it turned out, however, that the improvement resulted to a small extent only from the particular sequence and more so from an overall improvement of the apparatus. We describe here the properties of the ideal MREV eight-pulse cycle insofar as they differ from those of the WAHUHA cycle.

a. Properties Related to the Detection of the NMR Signal

The initial amplitude of the oscillating part of the NMR signal can be made equal to the initial amplitude of a free-induction decay signal by applying a properly chosen 90° preparation pulse (see Fig. 5-1). The maximum potential signal amplitude can thus be exploited fully with ease. The oscillating signal does not ride on a dc pedestal as it does with the WAHUHA sequence. The absence of the pedestal is of particular importance when spin–lattice relaxation and/or (weak) I–S dipolar couplings cause in WAHUHA experiments a decay of the pedestal, which is considerably inconvenient for the subsequent processing of the data.

b. Properties of Zeroth-Order Average Hamiltonian

First-rank I-spin interactions are retained, but scaled down by a factor of $\sqrt{2/3}$, which is smaller than the corresponding WAHUHA scaling factor.

c. First-Order Corrections

The total MREV eight-pulse cycle is not symmetric; hence, $\mathscr{H}^{(1)}$ does not vanish identically. With the aid of Fig. 5-3 it is easily verified that the offset ($\Delta\omega$) and chemical shift ($\Delta\omega_i$) first-order correction terms are given by

$$\mathscr{H}_{\text{off}}^{(1)} + \mathscr{H}_{\text{CS}}^{(1)} = -\tfrac{1}{3}\tau \sum_i (\Delta\omega + \Delta\omega_i)^2 (I_x{}^i - I_z{}^i). \tag{5-4}$$

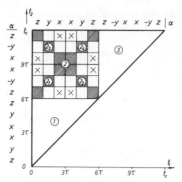

FIG. 5-3. First-order correction term $\tilde{\mathscr{H}}^{(1)}$ for MREV eight-pulse cycle.

$$\tilde{\mathscr{H}}_{CS}(t) + \tilde{\mathscr{H}}_{off}(t) \propto I_\alpha.$$

Domains ① and ③ encompass *symmetric* subcycles and do not contribute to $\tilde{\mathscr{H}}^{(1)}$. Domains $2_1 - 2_4$ are equivalent. $\tilde{\mathscr{H}}(t_2)$ and $\tilde{\mathscr{H}}(t_1)$ commute in cross-hatched areas. Contributions from areas marked by crosses cancel.

The first-order dipolar correction term vanishes: Domains ① and ③ in Fig. 5-3 encompass symmetric subcycles that do not contribute to $\tilde{\mathscr{H}}^{(1)}$, and domain ② involves the sum $\mathscr{H}_D{}^x + \mathscr{H}_D{}^y + \mathscr{H}_D{}^z = 0$. The same arguments imply that the first-order dipolar-offset and dipolar-chemical-shift cross terms vanish.

d. *Second-Order Corrections*

The purely dipolar second-order correction term is identical with the corresponding WAHUHA term if it is written in terms of τ instead of t_c. There are again cross terms between $\mathscr{H}_D$, $\mathscr{H}_{CS}$, $\mathscr{H}_{off}$, and $\mathscr{H}_j$. The coefficients are of the same order of magnitude as for the WAHUHA sequence.

From this enumeration the following conclusions may be drawn:

(i) The ideal (later) MREV cycle is hardly superior to the ideal (earlier) WAHUHA cycle. In Section D, where we consider pulse imperfections, it will become obvious that the MREV cycle is superior to the WAHUHA cycle.

(ii) The resolution "should" be best close to resonance since the second-order cross terms between $\mathscr{H}_D$ and $\mathscr{H}_{off}$ vanish at, and are very small close to resonance.

(iii) $\tilde{\mathscr{H}}_D^{(2)}$ appears to be the resolution-limiting factor.

Experimentally it is found in agreement with (ii) that the resolution deteriorates far off resonance[89,90]; however, since our early multiple-pulse experiments

[89] W. K. Rhim, D. D. Elleman, and R. W. Vaughan, *J. Chem. Phys.* **59**, 3740 (1973).
[90] A. N. Garroway, P. Mansfield, and D. C. Stalker, *Phys. Rev. B* **11**, 121 (1975).

on CaF_2 [46,51] it became clear that the resolution tends consistently to be better somewhere off, rather than exactly on resonance. Later we realized that this is due to a further averaging process, which becomes operative off resonance and which we are going to describe now.

B. Off-Resonance Averaging

We start by recalling that the suppression of homonuclear dipolar couplings discussed so far is the combined result of two averaging processes:

(i) The application of a strong magnetic field B_{st} imposes a fast common motion on the spins, which makes it possible—even mandatory—to truncate the full dipolar Hamiltonian. We are left with the truncated or secular dipolar Hamiltonian, which is just the time average of the time-dependent full dipolar Hamiltonian in the rotating frame. There was and is no need at this stage to consider corrections $\mathcal{H}^{(1)}$, $\mathcal{H}^{(2)}$, etc., to the average Hamiltonian because the Larmor period—which plays the role of the cycle time—is typically smaller by three orders of magnitude than $T_{2,\text{rigid}}$ in very low, in principle in zero, applied field.

(ii) The application of any one of the line-narrowing multiple-pulse sequences swirls all spins synchronously around and leads to a zero average of the truncated dipolar couplings in the toggling frame. Because the cycle time of the multiple-pulse sequence typically cannot be made much smaller than $T_{2,\text{rigid}}$ in high fields we had to consider correction terms to the average Hamiltonian. The lowest order nonvanishing purely dipolar correction term is $\mathcal{H}_D^{(2)}$ for both the WAHUHA and MREV cycles.

The effective Hamiltonian in the toggling frame with the applied field set somewhat off resonance consists of

(i) a zeroth-order resonance-offset term, the detailed structure of which depends on the particular pulse sequence at hand,

(ii) $\mathcal{H}_D^{(2)}$,

(iii) a host of cross terms,

(iv) a host of pulse imperfection terms.

All these terms have equal rights in the effective Hamiltonian irrespective of whether they arise just as averages or as higher order correction terms.

In the toggling frame the resonance-offset term imposes a common uniform motion on all spins of a given isotopic species. It is just a third repetition of our by now familiar game to treat this common motion of the spins by a further interaction representation, U_{off}. The consequence is that the remaining parts of the toggling-frame effective Hamiltonian—$\bar{\mathscr{H}}_D^{(2)}$, cross terms, pulse imperfection terms—acquire a time dependence, e.g.,

$$\bar{\mathscr{H}}_D^{(2)} \rightarrow \bar{\mathscr{H}}_D^{(2)}(t) = U_{\text{off}}^{-1}(t)\,\bar{\mathscr{H}}_D^{(2)}U_{\text{off}}(t).$$

Over this time dependence we may average eventually—provided it is fast enough.

U_{off} represents a rotation

(i) about the toggling-frame 111 direction for the WAHUHA four-pulse sequence,

(ii) about the toggling-frame 101 direction for the MREV eight-pulse sequence.

It is over these motions that we must average the time-dependent toggling-frame Hamiltonian. (Recall that the toggling and rotating frames can be considered to coincide if the spin system is observed stroboscopically.)

What is the result of this third averaging process? First of all, the time average of $\bar{\mathscr{H}}_D^{(2)}(t)$ vanishes for the WAHUHA four-pulse experiment! It is possible, but rather tedious, to demonstrate this for a continuous rotation about the 111 toggling-frame axis (see Appendix C). It is much easier—and still sufficiently instructive—to consider a stepwise rotation with three steps for one full turn. This we shall do now.

We expressed $\bar{\mathscr{H}}_D^{(2)}$ in terms of $\mathscr{H}_D{}^x$, $\mathscr{H}_D{}^y$, and $\mathscr{H}_D{}^z$. The structure of $\mathscr{H}_D{}^x$ etc., is evidently such that each 120° rotational step about the 111 axis carries successively

$$x \rightarrow y, \qquad y \rightarrow z, \qquad z \rightarrow x.$$

The averge of $\bar{\mathscr{H}}_D^{(2)}(t)$ becomes

$$
\begin{aligned}
\langle \bar{\mathscr{H}}_D^{(2)}(t)\rangle_{\text{av}} = \frac{1}{3}\frac{t_c{}^2}{648}\{&[(\mathscr{H}_D{}^x - \mathscr{H}_D{}^z), [\mathscr{H}_D{}^x, \mathscr{H}_D{}^y]] \\
+ \quad &[(\mathscr{H}_D{}^y - \mathscr{H}_D{}^x), [\mathscr{H}_D{}^y, \mathscr{H}_D{}^z]] \\
+ \quad &[(\mathscr{H}_D{}^z - \mathscr{H}_D{}^y), [\mathscr{H}_D{}^z, \mathscr{H}_D{}^x]]\}. \quad (5\text{-}5)
\end{aligned}
$$

A consequence of $\mathscr{H}_D{}^x + \mathscr{H}_D{}^y + \mathscr{H}_D{}^z = 0$ is

$$[\mathscr{H}_D{}^x, \mathscr{H}_D{}^y] = [\mathscr{H}_D{}^y, \mathscr{H}_D{}^z] = [\mathscr{H}_D{}^z, \mathscr{H}_D{}^x].$$

The inner commutators in Eq. (5-5) can therefore be bracketed out and the terms of the remaining factor cancel. Hence we have the result that the average of $\bar{\mathscr{H}}_D^{(2)}(t)$ vanishes for the WAHUHA sequence.

Off-resonance averaging substantially reduces but does not eliminate completely $\bar{\mathscr{H}}_D^{(2)}$ for the MREV eight-pulse sequence. The direction of the off-resonance rotation axis is not as favorable as for the four-pulse sequence.

Let us ask now: How important is off-resonance averaging in practice? This question is, in fact, somewhat too general. Therefore, let us ask in more detail:

(i) What about experiments that clearly demonstrate the effectiveness of off-resonance averaging?

(ii) What consequence has off-resonance averaging for the design of highly efficient line-narrowing multiple-pulse sequences?

(i) Simple, rather than sophisticated, multiple-pulse sequences are best suited for demonstrating the effectiveness of off-resonance averaging. An example is the phase alternated tetrahedral angle (PAT) sequence[91,92] sketched in Fig. 5-4.

On resonance, the PAT sequence does not average out the dipolar couplings between I spins. In fact, no pulse sequence consisting of two rf pulses only can do that.[51] Indeed, in an experiment on resonance on CaF_2, we could lengthen the decay of the ^{19}F NMR signal barely by a factor of two.

Off-resonance averaging theory predicts a complete suppression of the I-spin dipolar couplings,[91] and indeed a more than threefold stretching of the decay—as compared with the on-resonance case—was observed on going off resonance by 8 kHz.[91]

Pines and Waugh[93] realized that the stretching of the decay in this particular experiment was limited by a finite pulse-width effect, and that it could be overcome by adapting the nutation angle to the duty factor $2t_W/t_c$ of the sequence. These authors report having attained with the PAT sequence a ^{19}F decay time in CaF_2 in excess of 1 msec. They also give further examples of simple pulse sequences with marked off-resonance averaging effects together with a very intricate theory.

(ii) While there is no doubt that off-resonance averaging helps improve resolution with WAHUHA four- and MREV eight-pulse sequences there is

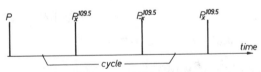

FIG. 5-4. The phase alternated tetrahedral angle (PAT) sequence. P is a preparation pulse and may be just another $P_{-x}^{109.5}$ pulse.

[91] U. Haeberlen, J. D. Ellett, and J. S. Waugh, *J. Chem. Phys.* **55**, 53 (1971).

[92] J. D. Ellett, Ph.D. Thesis, Massachusetts Institute of Technology, Cambridge, 1970 (unpublished).

[93] A. Pines and J. S. Waugh, *J. Magn. Resonance* **8**, 354 (1972).

doubt that for the four-pulse sequence the improvement results predominantly from the suppression of $\mathscr{H}_D^{(2)}$. We are led to this suspicion by the fact that the MREV eight-pulse sequence—where $\mathscr{H}_D^{(2)}$ is not fully eliminated by off-resonance averaging—seems to yield significantly better resolution than the four-pulse sequence, where $\mathscr{H}_D^{(2)}$ is fully eliminated.

A side conclusion from this observation is that there is no practical point now to use sophisticated multiple-pulse sequences for which not only $\mathscr{H}_D = \mathscr{H}_D^{(1)} = 0$, but also $\mathscr{H}_D^{(2)} = 0$. The first such sequence has been proposed by Evans,[94] and Mansfield[83] has devised a whole family of sequences that possess this property.

So, what gets eliminated by off-resonance averaging in WAHUHA and, above all, in MREV experiments? We get a hint by noting that the amount by which the field has to be set off resonance in order to attain the best resolution tends to become smaller and smaller as experimenters improve their equipment more and more. This indicates that off-resonance averaging is also effective in suppressing adverse pulse imperfection effects. With "poor" equipment this is probably the most important off-resonance effect in MREV and WAHUHA experiments.

We shall study pulse imperfections rather carefully in Section D, but first we shall consider another interesting resonance offset property of multiple-pulse sequences, namely, a dramatic difference in relaxation rates parallel and perpendicular to the effective field in the toggling frame.

C. $T_\parallel$ and $T_\perp$; Heteronuclear Dipolar Couplings

Consider Fig. 5-5, which is taken from Haeberlen et al.[91] It shows the ^{19}F NMR signal of CaF_2 in an on- and an off-resonance WAHUHA experiment.

On resonance, the magnetization $\langle \mathbf{M} \rangle$ decays nearly to zero over the time of the WAHUHA experiment, off resonance it does not. The oscillations almost do, but not the pedestals. As we have seen in Chapter IV, the pedestals arise from the component of $\langle \mathbf{M} \rangle$ that is initially parallel to $\mathbf{B}_{eff}$ in the toggling frame, whereas the oscillations arise from the component of $\langle \mathbf{M} \rangle$ that is perpendicular to $\mathbf{B}_{eff}$. For convenience we shall characterize by time constants $T_\perp$ and $T_\parallel$ the decays of $\langle \mathbf{M} \rangle$, respectively, perpendicular and parallel to $\mathbf{B}_{eff}$, even though these decays need not be simple exponential. $T_\perp$ corresponds to the usual WAHUHA decay time. That two relaxation rates are involved in NMR line-narrowing experiments had already become evident from the work

[94] W. A. B. Evans, private communication (see Haeberlen and Waugh[51]).

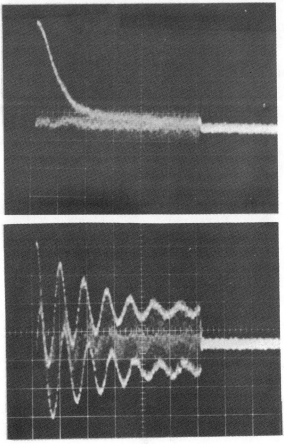

FIG. 5-5. WAHUHA experiment performed on a single crystal of CaF$_2$. Horizontal scales = 1.0 msec/division. Upper trace, on resonance; lower trace, 2.0 kHz off resonance (from Haeberlen et al.[91]).

of Lee and Goldburg[61] and Tse and Hartmann.[95] In both of their experiments solid samples were irradiated with a strong rf field whose intensity and frequency offset from resonance were chosen to produce an effective field in the rotating coordinate frame, which made the magic angle with the z axis. Lee and Goldburg measured the rate of decay ($T_\perp^{-1}$) of the component of $\langle \mathbf{M} \rangle$ perpendicular to the effective field, while Tse and Hartmann studied the relaxation ($T_\parallel$) of the magnetization along $\mathbf{B}_{\text{eff}}$ and they found $T_\parallel$ to be orders of magnitude longer than $T_\perp$ as found by Lee and Goldburg.

[95] D. Tse and S. R. Hartmann, *Phys. Rev. Lett.* **21**, 511 (1968).

In one experiment we measured $T_{\parallel}$ in an off-resonance WAHUHA experiment on a single crystal of CaF_2 doped with paramagnetic U^{4+} ions and found it to exceed 0.6 sec, which is two orders of magnitude greater than its on-resonance value. (A $T_{\parallel}$ of 0.6 sec exceeds the spin–lattice relaxation time T_1 of that crystal. This is possible because in the WAHUHA experiment, as in the experiment of Tse and Hartmann, the spin diffusion relaxation mechanism is suppressed.) $T_{\perp}$ only increased by a factor of roughly three on going the same amount off resonance.

We can use the resonance offset-averaging theory to understand in a qualitative way the difference between the resonance offset dependences of $T_{\parallel}$ and $T_{\perp}$. The on-resonance decay of the magnetization is determined by $\mathscr{H}_D^{(2)}$ and pulse imperfections. We have seen above that off resonance we must average these correction terms over the motion caused by $\mathbf{B}_{\text{eff}}$ in the toggling frame. As in Chapter IV, Section C,4,d, it is again helpful to introduce a new set of axes in the toggling frame, the Z axis of which points along the (old) 111 direction. The average of a typical correction term $\mathscr{H}_{\lambda}^{(n)}$ over the motion generated by $\mathbf{B}_{\text{eff}}$ has the form

$$\langle \mathscr{H}_{\lambda}^{(n)} \rangle_{\text{av}} = \frac{1}{2\pi} \int_0^{2\pi} \exp(i\Phi I_Z)\, \mathscr{H}_{\lambda}^{(n)} \exp(-i\Phi I_Z)\, d\Phi.$$

This expression commutes with I_Z, which we can easily see if we express $\mathscr{H}_{\lambda}^{(n)}$ in terms of spherical tensor operators T_{lm} to take advantage of their simple transformation and commutation properties.
$\langle \mathscr{H}_{\lambda}^{(n)} \rangle_{\text{av}}$ can be written as a sum of terms of the form

$$\frac{1}{2\pi} \int_0^{2\pi} \exp(i\Phi I_Z) T_{lm} \exp(-i\Phi I_Z)\, d\Phi = \frac{1}{2\pi} \int_0^{2\pi} \exp(im\Phi) T_{lm}\, d\Phi = \delta_{m0}\, T_{l0}.$$

But $[I_Z, T_{l0}] = 0$.

Therefore, back in the old toggling frame (or rotating frame if observation of the spin system is restricted to the proper windows) we have

$$[\langle \mathscr{H}_{\lambda}^{(n)} \rangle_{\text{av}}, (I_x + I_y + I_z)] = 0.$$

Thus, for a sufficiently large resonance offset the correction terms to the on-resonance average Hamiltonian are replaced by terms that commute with $(I_x + I_y + I_z)$. It follows immediately from this and the equation of motion (Heisenberg equation) for the magnetization operator $\mathbf{M}(t)$,

$$\dot{\mathbf{M}}(t) = i\left[\left(\sum_{\lambda}\sum_{n} \langle \mathscr{H}_{\lambda}^{(n)} \rangle_{\text{av}} + \cdots \right), \mathbf{M}(t)\right],$$

that the time derivative of the component of the magnetization parallel to the 111 direction in the toggling frame is zero. [The dots indicate higher order corrections to the present averages.] This accounts for the fact that in multiple-pulse line-narrowing experiments performed off resonance the components of $\langle \mathbf{M} \rangle$ parallel to the 111 direction decay much more slowly than the components of $\langle \mathbf{M} \rangle$ perpendicular to that direction.

On resonance, $T_\perp$ is determined by a number of correction terms, some of which commute with $(I_x + I_y + I_z)$ and some of which do not. Off resonance, only terms that commute with $(I_x + I_z + I_y)$ remain to first order, so that $T_\perp$ increases somewhat off resonance, but not so much as $T_\parallel$, which is not affected at all by these terms. Since the average chemical-shift Hamiltonian $\mathscr{H}_{CS}$ commutes with $(I_x + I_y + I_z)$, the Heisenberg equation of motion above implies that the component of $\langle \mathbf{M} \rangle$ along the 111 direction will not carry information about the chemical shifts of the sample. Thus we see that $T_\perp$, rather than $T_\parallel$, gives a measure of the chemical shift resolution of a multiple-pulse line-narrowing experiment.

In crystals containing more than one species of nuclear spins the WAHUHA $T_\parallel$ is much greater than $T_\perp$ both on and off resonance. Figure 5-6 (again from Haeberlen et al.[91]) shows the WAHUHA ^{19}F decay of a single crystal of LaF$_2$. Notice that there is a rapid initial decay of the signal train to $\approx \frac{1}{3}$ of its initial value, followed by a much slower decay. The initial decay time is 48 μsec, which is longer than the 20.2 μsec ^{19}F free-induction decay time of the crystal at this orientation. The slowly decaying signal never develops a beat structure as the spectrometer frequency is shifted away from resonance, although the time constant of the decay decreases as the resonance offset is increased beyond several kilohertz. The difference between $T_\parallel$ and $T_\perp$ in this experiment follows from the nature of the truncated heteronuclear dipolar interaction Hamiltonian, $\mathscr{H}^{D,IS}_{secular}$, which in this experiment dominated the decay.[96] It behaves in the I-spin space as a first-rank tensor so its average during the pulse sequence, both on and off resonance is

$$\mathscr{H}^{D,IS}_{secular} = \tfrac{1}{3}\gamma_n{}^I \gamma_n{}^S \sum_{i,k} r_{ik}^{-3}(1 - 3\cos^2 \vartheta_{ik})\, S_z{}^i (I_x{}^k + I_y{}^k + I_z{}^k).$$

As usual, the I-spin species (^{19}F) is the one we observe and hit by the multiple-pulse sequence. This average Hamiltonian commutes with $(I_x + I_y + I_z)$ and as noted earlier, this implies that the decay of the component of $\langle \mathbf{M} \rangle$ along the 111 direction will be slower than the decay of the component of $\langle \mathbf{M} \rangle$ perpendicular to that direction. Thus $T_\parallel$ should exceed $T_\perp$, as indeed is the case in Fig. 5-6. The resonance offset average Hamiltonian commutes with $(I_x + I_y + I_z)$, which accounts for the fact that no beat structure was observed on the slowly decaying signal.

---

[96] Suppression of $\mathscr{H}^{D,IS}_{secular}$ by either heteronuclear or self-decoupling has been discussed in Chapter IV, Section F, 5.

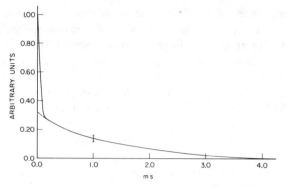

FIG. 5-6. WAHUHA experiment performed on a single crystal of LaF_2 (from Haeberlen et al.[91]).

D. Pulse Imperfections

Naturally we can never excite our spin systems with ideal pulses in real experiments. Since the averaging achievable with ideal pulse sequences—which we have considered thus far—appears to be fantastically good, and since the results of the pioneering experiments in all multiple-pulse laboratories were or are not that good, suspicion has arisen that the discrepancy is due to pulse imperfections. We are aware of a host of pulse imperfections and there may be more we are not aware of. For most of the pulse imperfections that we understand, compensation schemes have been worked out. We shall discuss them in this section.

The following pulse imperfections have attracted interest thus far:

finite pulse widths,
flip angle errors common to all pulses,
rf inhomogeneity,
power droop,
flip angle errors different for the different sets of pulses,
phase errors,
phase transients.

We shall consider here another conceivable imperfection, namely, timing errors.

1. Finite Pulse Widths; Common Flip Angle Errors; rf Inhomogeneity; Power Droop

These pulse imperfections form a class to which we shall turn first. Their important common feature is that they do not destroy the cyclic and the symmetry properties of multiple-pulse sequences (see below). For a power droop this is true only in so far as the power droop over individual cycles can be neglected. Our approach will be completely straightforward: We consider the average Hamiltonian of spin systems subject to multiple-pulse sequences composed of pulses of finite width that flip the nuclear magnetization through variable angles β. We shall learn how the zeroth-order effects of finite pulse widths can be overcome.[97] We shall further learn to understand the sensitivity of various pulse sequences to flip angle errors common to all pulses and thus also to a droop of the rf power in the course of a long pulse train. Our approach also provides an understanding for the effects of rf inhomogeneity since rf inhomogeneity means nothing but a distribution over the sample volume of common flip angle errors. Again we shall start by considering the WAHUHA cycle. The (superior) properties of the MREV eight-pulse cycle can easily be inferred from those of the WAHUHA four-pulse cycle.

As this class of pulse errors does not destroy the symmetry properties of multiple-pulse sequences, the WAHUHA sequence remains symmetric even when the pulses have a finite width. As a result all corrections of odd order to the average Hamiltonian still vanish automatically (see Chapter IV, Section D,4). We think it is reasonable to suspect that second-order corrections, and higher order corrections of even order in general, are modified only slightly by pulse imperfections—which we are trying to keep down anyhow—so it is not worthwhile going through the trouble of studying them for imperfect pulse cycles.

a. *WAHUHA Four-Pulse Cycle with Pulses of Finite Width;*
   Condition for $\mathcal{H}_{\mathrm{D}} = 0$; Scaling Factor

Figure 4-2 shows the timing of a WAHUHA sequence with pulses of finite width. Also shown is the evolution of the propagation operator $U_{\mathrm{rf}}(t)$, which is evidently symmetric about the midpoint of the second large window. The last two columns of Fig. 4-2 are intended to provide a "look and see" proof of that fact. Since $U_{\mathrm{rf}}(t)$ is symmetric, $\mathcal{H}_T^{\mathrm{int}}(t) = U_{\mathrm{rf}}^{-1}(t)\,\mathcal{H}_{\mathrm{secular}}^{\mathrm{int}}\,U_{\mathrm{rf}}(t)$ is also symmetric. The consequence of this property of $\mathcal{H}_T^{\mathrm{int}}(t)$ with regard to $\mathcal{H}^{(1)}$ has been mentioned above. Another consequence is that for evaluating $\mathcal{H}$ it is sufficient to consider only one half of the cycle. The average of $\mathcal{H}_{\mathrm{secular}}^{\mathrm{D}}(t)$, in particular, is proportional to

$$\mathbf{I}^i \cdot \mathbf{I}^k - 3\langle I_z^{\,i}(t)\, I_z^{\,k}(t)\rangle_{\mathrm{av}} = \mathbf{I}^i \cdot \mathbf{I}^k - \frac{1}{3\tau} \sum_{p=1}^{5} 3\langle I_z^{\,i}(t)\, I_z^{\,k}(t)\rangle_p\, t_p$$

where the summation is over the intervals $p = 1,\ldots,5$ of the first half of the WAHUHA cycle; $\langle\cdots\rangle_p$ means average over interval p; t_p is the duration of interval p.

[97] M. Mehring, *Z. Naturforsch. A* **27**, 1634 (1972).

TABLE 5-1

AVERAGES OF $I_z^i(t)I_z^k(t)$ OVER INTERVALS $p=1,\ldots,5$ OF WAHUHA SEQUENCE (SEE FIG. 4-2)[a]

p	Coefficient of					
	$I_x^iI_x^k$	$I_y^iI_y^k$	$I_z^iI_z^k$	$\frac{1}{2}(I_x^iI_y^k+I_y^iI_x^k)$	$\frac{1}{2}(I_x^iI_z^k+I_z^iI_x^k)$	$\frac{1}{2}(I_y^iI_z^k+I_z^iI_y^k)$
1	—	—	$\tau\left(1-\dfrac{t_w}{2\tau}\right)$	—	—	—
2	—	$\dfrac{t_w}{2}\left(1-\dfrac{\sin 2\beta}{2\beta}\right)$	$\dfrac{t_w}{2}\left(1+\dfrac{\sin 2\beta}{2\beta}\right)$	—	—	$t_w\dfrac{\sin^2\beta}{\beta}$
3	—	$\tau\left(1-\dfrac{t_w}{\tau}\right)\sin^2\beta$	$\tau\left(1-\dfrac{t_w}{\tau}\right)\cos^2\beta$	—	—	$\tau\left(1-\dfrac{t_w}{\tau}\right)\sin 2\beta$
4	$\dfrac{t_w}{2}\left(1-\dfrac{\sin 2\beta}{2\beta}\right)$	$\dfrac{t_w}{2}\left(1+\dfrac{\sin 2\beta}{2\beta}\right)\sin^2\beta$	$\dfrac{t_w}{2}\left(1+\dfrac{\sin 2\beta}{2\beta}\right)\cos^2\beta$	$t_w\dfrac{\sin^3\beta}{\beta}$	$t_w\dfrac{\sin^2\beta}{\beta}\cos\beta$	$\dfrac{t_w}{2}\left(1+\dfrac{\sin 2\beta}{2\beta}\right)\sin 2\beta$
5	$\tau\left(1-\dfrac{t_w}{2\tau}\right)\sin^2\beta$	$\tau\left(1-\dfrac{t_w}{2\tau}\right)\cos^2\beta\sin^2\beta$	$\tau\left(1-\dfrac{t_w}{2\tau}\right)\cos^4\beta$	$\tau\left(1-\dfrac{t_w}{2\tau}\right)\sin 2\beta\sin\beta$	$\tau\left(1-\dfrac{t_w}{2\tau}\right)\sin 2\beta\cos\beta$	$\tau\left(1-\dfrac{t_w}{2\tau}\right)\sin 2\beta\cos^2\beta$

[a] The coefficients of the various operator forms of $I_z^i(t)I_z^k(t)$ multiplied by t_p are shown. t_p is the length of interval p.

It is straightforward to evaluate the averages $\langle I_z^{\,i}(t)\, I_z^{\,k}(t)\rangle_p$. The heading in Table 5-1 shows which operator forms are generated by $U_{\mathrm{rf}}(t)$ from $I_z^{\,i}I_z^{\,k}$. The further entries of the table show the coefficients, multiplied by t_p, of these operators in the averages $\langle I_z^{\,i}(t)\, I_z^{\,k}(t)\rangle_p$.

The average of $\mathscr{H}_{\mathrm{secular}}^{\mathrm{D}}(t)$ contains, of course, the same spin operator combinations as does the heading in Table 5-1. The coefficients of these operators in the average of $\mathscr{H}_{\mathrm{secular}}^{\mathrm{D}}(t)$ are given in Table 5-2. Starting from Table 5-1, only simple juggling with trigonometric functions is required to find the expressions of Table 5-2.

The outstanding feature of Table 5-2 is that all the coefficients have one factor in common. Suppression of dipolar interactions is evidently obtained when this factor becomes zero, which means when

$$\left[\left(1 - \frac{t_{\mathrm{w}}}{2\tau}\right)\cos\beta + \frac{t_{\mathrm{w}}}{2\tau}\frac{\sin\beta}{\beta}\right] = 0. \tag{5-6}$$

In Eq. (5-6) let us consider $t_{\mathrm{w}}/2\tau$ as a parameter and the flip angle $\beta = \omega_1 t_{\mathrm{w}}$ as a variable. Note that this assignment of roles follows closely at least a possible procedure of alignment of actual multiple-pulse sequences. The range of t_{w}/τ is clearly $0 \leqslant t_{\mathrm{w}}/\tau \leqslant 1$. The equality signs apply only for idealized cases. By a different approach, which did not state the problem as clearly as do Table 5-2 and Eq. (5-6), we recognized very early[51] that $\mathscr{H}_{\mathrm{secular}}^{\mathrm{D}}$ can be suppressed not only for $t_{\mathrm{w}} = 0$, but also for small but finite values of the parameter t_{w}/τ. However, it was Mehring[97] who realized—still using a somewhat different approach—that there is a choice of β for the entire range of

TABLE 5-2

Average of $\mathbf{I}^i \cdot \mathbf{I}^k - 3I_z^{\,i}I_z^{\,k}$ over WAHUHA Four-Pulse Cycle with Pulses of Finite Width $t_{\mathrm{w}}{}^a$

Operator	Coefficient
$I_x^{\,i}I_x^{\,k}$	$\left[\left(1 - \dfrac{t_{\mathrm{w}}}{2\tau}\right)\cos\beta + \dfrac{t_{\mathrm{w}}}{2\tau}\dfrac{\sin\beta}{\beta}\right]\cos\beta$
$I_y^{\,i}I_y^{\,k}$	$[\cdots]\quad \cos^3\beta$
$I_z^{\,i}I_z^{\,k}$	$-[\cdots](1+\cos^2\beta)\cos\beta$
$\frac{1}{2}(I_x^{\,i}I_y^{\,k}+I_y^{\,i}I_x^{\,k})$	$-[\cdots]\quad 2\sin^2\beta$
$\frac{1}{2}(I_x^{\,i}I_z^{\,k}+I_z^{\,i}I_x^{\,k})$	$-[\cdots]\quad 2\sin 2\beta$
$\frac{1}{2}(I_y^{\,i}I_z^{\,k}+I_z^{\,i}I_y^{\,k})$	$-[\cdots](1+\cos^2\beta)2\sin\beta$

^{a} The coefficients of the relevant spin operator combinations in $\langle \mathbf{I}^i \cdot \mathbf{I}^k - 3I_z^{\,i}I_z^{\,k}\rangle_{\mathrm{av}}$ are shown. The expression in the square brackets, $[\cdots]$, is the same in all coefficients.

t_W/τ that satisfies Eq. (5-6). We shall denote the particular value of the flip angle β that satisfies Eq. (5-6) by β_0, which is a function of t_W/τ.

For $t_W/\tau = 0$, β_0 is, of course, equal to $\frac{1}{2}\pi$. For $t_W/\tau = 1$, Eq. (5-6) leads to the transcendental equation $\tan\beta_0 = -\beta_0$, which has $\beta_0 = 116.24°$ as a solution. For intermediate values of t_W/τ, β_0 varies monotonically between these two limiting values. Mehring[97] gives a plot of β_0 versus $\frac{2}{3}t_W/\tau$ (which is the duty factor of the pulse train).

In our experiments we usually select a certain value for t_W/τ, typically 0.2. Then we adjust β for optimum resolution. If everything is aligned properly we indeed obtain the optimum resolution for β slightly larger than $\frac{1}{2}\pi$.

In summary, homonuclear dipolar interactions can be suppressed up to and including first-order corrections in the Magnus expansion by WAHUHA four-pulse sequences with pulses of finite width. The condition is that the flip angle β is chosen such that it satisfies Eq. (5-6) for the selected ratio of t_W/τ.

Finite pulse widths affect not only the suppression of dipolar interactions, but also the scaling factor and, weakly, the off-resonance averaging mechanism. To study their influence on the scaling factor we must consider the average of $I_z(t)$. This is easily obtained with the aid of Fig. 4-2. We give only the result:

$$
\begin{aligned}
\langle I_z(t)\rangle_{av} \\
= \langle U_{rf}^{-1}(t)\, I_z\, U_{rf}(t)\rangle_{av} \\
= \frac{1}{3}\Bigg\{ I_x\Bigg[\sin\beta - \frac{t_W}{\tau}\left(\frac{\sin\beta}{2} - \frac{1-\cos\beta}{\beta}\right)\Bigg] \\
+ I_y\Bigg[\sin\beta(1+\cos\beta) - \frac{t_W}{\tau}\left(\sin\beta\left(1 + \frac{\cos\beta}{2}\right) - \frac{1+\sin^2\beta-\cos\beta}{\beta}\right)\Bigg] \\
+ I_z\Bigg[(1+\cos\beta+\cos^2\beta) \\
- \frac{t_W}{\tau}\left(\cos\beta + \frac{1+\cos^2\beta}{2} - (1+\cos\beta)\frac{\sin\beta}{\beta}\right)\Bigg]\Bigg\}.
\end{aligned}
\tag{5-7}
$$

Denoting the coefficients of I_x, I_y, and I_z in Eq. (5-7) by C_x, C_y, and C_z, respectively, the scaling factor S may be expressed by

$$
S = \frac{1}{3}(C_x^2 + C_y^2 + C_z^2)^{1/2}.
\tag{5-8}
$$

With the aid of any decent pocket-sized calculator it is easy to verify that S varies monotonically between $3^{-1/2} = 0.57735$ and 0.56606 as t_W/τ increases from 0 to 1, provided β is always chosen such that it satisfies Eq. (5-6). This very small variation of the scaling factor S is of no practical importance, particularly as S is an easily measurable quantity. We are, of course, very glad that in return for being able to work with increased pulse widths we do not have to pay a high price in terms of an appreciably reduced scaling factor.

b. *Flip Angle Errors Common to All Pulses; rf Inhomogeneity; Power Droop*

State-of-the-art multiple-pulse spectrometers usually have a means to monitor the pulse power of the transmitter. By turning on the corresponding knob one adjusts for the optimum flip angle β_0. Misadjustments lead to flip angle errors common to all pulses. Much more dangerous are, however, flip angle errors that are not under control of the spectroscopist. (We contrast the spectroscopist to the designer and builder of the spectrometer although often they are the same person.) Such errors are caused by, e.g., all short- and long-term variations and drifts of the transmitter power output. The notorious power droop—the decrease of the transmitter power output during the course of a long pulse train—is just a special case. One source of these troubles is drifting and humming of dc power supplies. Another important source of common flip angle errors is the unavoidable rf inhomogeneity, which makes it impossible to adjust β to its optimum value β_0 for all volume elements of the sample.

The discussion of common flip angle errors, rf inhomogeneity, and power droop is contained in a discussion of the dependence on β of the NMR response of a solid sample to the particular multiple-pulse sequence at hand. We shall study this dependence here for the WAHUHA sequence. Our main concern is, of course, spectral resolution. There are two main reasons why a deviation ε of β from its optimum value β_0 (by definition, $\varepsilon = \beta - \beta_0$) affects the resolution attainable in multiple-pulse experiments:

(i) $\varepsilon \neq 0$ leads to an incomplete suppression of dipolar interactions.

(ii) $\varepsilon \neq 0$ changes the scaling factor S, and this leads via the rf inhomogeneity to a deterioration of the spectral resolution.

There is also an indirect consequence of flip angle errors: The variation with β of the direction of $\mathbf{B}_{\text{eff}}$ in the toggling frame makes the resonance offset-averaging mechanism dependent on β. We only mention this effect but do not dwell upon it.

We are well prepared for our current subject: The dependence on ε of $\mathscr{H}_{\text{D}} \equiv \langle \mathscr{H}^{\text{D}}_{\text{secular}}(t) \rangle_{\text{av}}$ is contained in the coefficients of Table 5-2, and the dependence on β of the scaling factor or, even more important, of $\langle I_z(t) \rangle_{\text{av}} \propto \mathscr{H}_{\text{CS}} + \mathscr{H}_{\text{off}}$ is contained in Eq. (5-7). For better perception of the sensitivity to ε of $\mathscr{H}_{\text{D}}$, $\mathscr{H}_{\text{CS}} + \mathscr{H}_{\text{off}}$, and the scaling factor, we have compiled in Table 5-3 the coefficients of the spin operator combinations occurring in $\mathscr{H}_{\text{D}}$ and $\mathscr{H}_{\text{CS}} + \mathscr{H}_{\text{off}}$ in terms of ε for $t_{\text{W}}/\tau \to 0$.

TABLE 5-3

Sensitivity to ε of the Coefficients of the Spin Operators Occurring in $\mathcal{H}_D$ and $\mathcal{H}_{CS} + \mathcal{H}_{off}$[a]

Operator	Coefficient in $\mathcal{H}_D$ or $\mathcal{H}_{CS} + \mathcal{H}_{off}$	for $t_W/\tau \to 0$	
$I_x^i I_x^k$	$\sin^2\varepsilon$	$\to \varepsilon^2$	$+ \ O(\varepsilon^4)$
$I_y^i I_y^k$	$\sin^4\varepsilon$	$\to$	$O(\varepsilon^4)$
$I_z^i I_z^k$	$-\sin^2\varepsilon(1+\sin^2\varepsilon)$	$\to -\varepsilon^2$	$+ \ O(\varepsilon^4)$
$[I_x^i I_z^k]_+$	$-4\cos\varepsilon\sin^2\varepsilon$	$\to -4\varepsilon^2$	$+ \ O(\varepsilon^4)$
$[I_x^i I_y^k]_+$	$2\cos^2\varepsilon\sin\varepsilon$	$\to 2\varepsilon$	$+ \ O(\varepsilon^2)$
$[I_y^i I_z^k]_+$	$2(1+\sin^2\varepsilon)\sin\varepsilon\cos\varepsilon$	$\to 2\varepsilon$	$+ \ O(\varepsilon^3)$
I_x	$\cos\varepsilon$	$\to 1 - \varepsilon^2/2$	$+ \ O(\varepsilon^4)$
I_y	$\cos\varepsilon(1-\sin\varepsilon)$	$\to 1 - (\varepsilon+\frac{1}{2}\varepsilon^2)+$	$O(\varepsilon^3)$
I_z	$1 - \sin\varepsilon + \sin^2\varepsilon$	$\to 1 - (\varepsilon-\varepsilon^2)$	$+ \ O(\varepsilon^3)$

[a] $[I_p^i I_q^k]_+$ is an abbreviation for $\frac{1}{2}(I_p^i I_q^k + I_q^i I_p^k)$.

The coefficients of $[I_x^i I_y^k]_+$, $[I_y^i I_z^k]_+$, I_y, and I_z depend linearly on ε. It is via the corresponding terms in $\mathcal{H}_D$ and $\mathcal{H}_{CS} + \mathcal{H}_{off}$ that the inhomogeneity of the rf field, a power droop, and a misalignment of the transmitter power affect by far most strongly the spectral resolution in WAHUHA experiments.

Residual dipolar line broadening is expressed by the first two of these terms, whereas the latter two lead—among other things—to line broadening via the rf inhomogeneity. In the following subsection we shall see that by combining variants of the WAHUHA sequence it is possible to eliminate all coefficients of bilinear spin operators that are linear in I_y. These are (what luck!) exactly those coefficients that are linear in ε and that are, as a result, the most disturbing. Therefore, we shall not discuss residual dipolar line broadening any further for the WAHUHA sequence.

On the other hand, it seems impossible to eliminate by this technique the coefficients linear in ε of linear spin operators, and thus the strongest direct line-broadening effect arising from rf inhomogeneity.

Garroway et al.[90] have shown, however, that by combining WAHUHA-type cycles with cycles that contain 270° pulses in addition to 90° pulses it is possible also to eliminate these terms. While the scheme has been proven to work well for liquids it remains to be seen whether it is also useful for solids.

The direct rf inhomogeneity line-broadening mechanism in WAHUHA experiments works as follows: In an rf coil there always exists a distribution of the strength B_1 of the rf field, and consequently a distribution of β over the sample volume. Let us assume that this latter distribution is centered at β_0, that it is bell-shaped, and that its half-width at half-height is $\langle \varepsilon^2 \rangle^{1/2}$. Consider a narrow resonance line of a liquid sample, which in the ordinary NMR spectrum is shifted off resonance by $\Delta\omega$. The center of the line appears in the multiple-pulse spectrum at $\Delta\omega S(\beta_0)$. Spins that are flipped by angles β different from β_0 "appear" in the multiple-pulse spectrum at $\Delta\omega S(\beta)$. $S(\beta)$ is given by Eq. (5-8). By expressing S in terms of ε and assuming for simplicity $t_W/\tau \ll 1$, which means $\beta_0 \approx \frac{1}{2}\pi$, we obtain

$$\Delta\omega S(\beta) \to \Delta\omega S(\beta_0 + \varepsilon) = \Delta\omega S(\tfrac{1}{2}\pi)(1 + \tfrac{2}{3}\varepsilon) = \Delta\omega(1/\sqrt{3})(1 + \tfrac{2}{3}\varepsilon). \quad (5\text{-}9)$$

The distribution of the strength of the rf field is thus reflected directly in the multiple-pulse lineshape of a liquid sample. The half-width at half-height of the line is $\Delta\omega(1/\sqrt{3})\frac{2}{3}\langle\varepsilon^2\rangle^{1/2}$. Note, in particular, that the width increases linearly with the offset $\Delta\omega$. For WAHUHA experiments on liquids this is typically the dominant line-broadening effect. For solids it is one line-broadening mechanism among several others.

In order to get an idea of its practical importance let us choose $\langle\varepsilon^2\rangle^{1/2} = 0.052 \triangleq 3°$ and $\Delta\omega = 2\pi \times 2000 \ \sec^{-1}$. These values lead to a line with a full width at half-height of as much as $2\pi \times 80$ Hz. Both input numbers of our example are absolutely realistic: It requires special techniques and efforts to wind small rf coils that produce rf fields substantially more homogeneous over reasonably sized samples than specified by $\langle\varepsilon^2\rangle^{1/2} \triangleq 3°$, although values less than 1° have been reported.[89] $\Delta\omega = 2\pi \times 2000$ Hz is a natural value to choose for the center of proton spectra, which at $\omega_0 = 2\pi \times 90$ MHz have a typical spread of $2\pi \times 2250$ Hz $\triangleq 25$ ppm. For ^{19}F work even substantially larger off-resonance shifts are often required.

c. *Residual Dipolar Line Broadening; Compensation Schemes;*
   MREV Eight-Pulse Sequence

With regard to residual dipolar line broadening, we recognized those terms in Tables 5-2 and 5-3 as the most dangerous ones that are linear in ε. As we have shown in 1968[51] it is possible to design compensation schemes that eliminate these troubling terms. The most successful of them—up to this date—seems to be the MREV eight-pulse sequence (see Fig. 5-1), which consists of two subcycles: the first is a WAHUHA cycle; the second is again a WAHUHA cycle, but the P_x and P_{-x} pulses are interchanged. The propagation operator $U_{rf}(t)$ runs during the first MREV subcycle through exactly the same sequence of states as it does in a WAHUHA sequence. During the second subcycle $U_{rf}(t)$ runs again through the same sequence of states, but the sign of I_x is reversed everywhere.

Let us consider $\mathcal{H}_D$ for the MREV sequence. The spin operator combinations involved are the same as for the WAHUHA sequence (see Table 5-2). The coefficients are one-half the sum of the respective coefficients from each subcycle. For the first one they are evidently identical with the corresponding coefficients of the WAHUHA cycle (see Table 5-2). We leave it as an easy exercise for the reader to show that the same coefficients are obtained again for the second subcycle; however, the signs are reversed of all those coefficients belonging to operators that contain I_y linearly.

As a result, the MREV-coefficients of $[I_x{}^i I_y{}^k]_+$ and $[I_y{}^i I_z{}^k]_+$ vanish identically, that is, regardless of the particular value of β. A corresponding result is obtained for $\mathscr{H}_{CS}$ and $\mathscr{H}_{off}$: The coefficient of I_y vanishes identically.

These results have highly important consequences: All remaining co-efficients in $\mathscr{H}_D$ now have *two* factors in common, namely,

$$\left[\left(1 - \frac{t_W}{2\tau}\right)\cos\beta + \frac{t_W}{2\tau}\frac{1}{\beta}\sin\beta\right] \quad \text{and} \quad \cos\beta.$$

This means that there are now two possible choices of β for which $\mathscr{H}_D$ vanishes. One is the same as for the WAHUHA cycle [see Eq. (5-6)] and the other is $\beta = \frac{1}{2}\pi$. Both choices coincide for $t_W/\tau = 0$.

Figure 5-7 shows how the surviving coefficients in $\mathscr{H}_D$ vary with β for three choices of t_W/τ. The interesting point to note is that they all stay very small—below 1%, say—for an appreciable range of β. This is true, in particular, when t_W/τ is, on the one hand, nonzero so that the two zeros of $\mathscr{H}_D$ are well separated, but when, on the other hand, it is small enough for the coefficients to remain negligibly small between the zeros. Figure 5-7c shows that this is definitely no longer the case for large pulse widths approaching the limiting case $t_W \to \tau$. It is therefore not advisable to operate MREV sequences with pulse widths approaching τ.

In summary, we may say that for small (but nonzero) pulse widths the suppression of dipolar spin–spin couplings by the MREV eight-pulse sequence is very insensitive to flip angle errors common to all pulses and therefore to misalignments, fluctuations, drifts, and a droop of the rf pulse power, and to the inhomogeneity of the rf field.

These properties of the MREV eight-pulse cycle—first predicted theoretically by Mansfield[83]—have been confirmed by specific experiments[55] and are confirmed by daily work in multiple-pulse laboratories—including ours—all over the world.

What about the direct rf inhomogeneity line-broadening mechanism in MREV experiments? For the WAHUHA sequence the coefficients in $\mathscr{H}_{CS}$ and $\mathscr{H}_{off}$ of both I_y and I_z are linear in ε. Only the coefficient of I_y is thrown out with the MREV sequence, so that we are left with one coefficient linear in ε. The counterpart of Eq. (5-9) is for the MREV sequence

$$\Delta\omega S(\beta) \to \Delta\omega S(\beta_0 + \varepsilon) \overset{\text{for } t_W/\tau \to 0}{=} \Delta\omega S(\tfrac{1}{2}\pi)(1 + \tfrac{1}{2}\varepsilon) = \Delta\omega\tfrac{1}{3}\sqrt{2}(1 + \tfrac{1}{2}\varepsilon).$$

$$(5\text{-}10)$$

By comparing Eqs. (5-9) and (5-10) we see that the direct rf inhomogeneity line-broadening mechanism is almost as effective for the MREV as for the

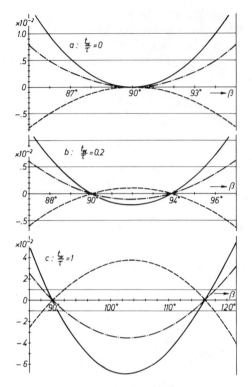

FIG. 5-7. Average dipolar Hamiltonian $\mathscr{H}_D$ for MREV sequence versus flip angle β. The curves show the coefficients of $\frac{1}{2}(I_x{}^iI_z{}^k+I_z{}^iI_x{}^k)$, ——; $I_x{}^iI_x{}^k$, — — —; and $I_z{}^iI_z{}^k$, —·—; for three different choices of t_W/τ. The coefficient of $I_y{}^iI_y{}^k$, though nonzero for $\beta \neq \frac{1}{2}\pi$ and β_0, is always negligibly small. Note that the horizontal and vertical scales, respectively, are equal for $t_W/\tau = 0$ and 0.2, but much larger for $t_W/\tau = 1$. Note in particular the range of β for which all coefficients stay smaller than, e.g., 10^{-2}. For $t_W/\tau = 0.2$ this range exceeds 9°.

WAHUHA sequence. Carefully optimizing the homogeneity of the rf field is therefore mandatory if one wishes to exploit fully the capability of suppressing dipolar spin–spin interactions in solids with any of these multiple-pulse sequences.

2. FLIP ANGLE AND PHASE ERRORS OF INDIVIDUAL PULSES; PHASE TRANSIENTS

We shall discuss this class of pulse imperfections together. Their common characteristic feature is that they destroy the cyclic property of multiple-pulse sequences. This means the propagation operator $U_{rf}(t)$ does not return to unity after a full cycle in the presence of these imperfections. While this seems disastrous for at least our description of multiple-pulse sequences it is, in fact, rather harmless. The following discussion will show how we can save the concept of cycles even in the presence of such pulse imperfections. Another of their common features is to render otherwise symmetric cycles nonsymmetric. So, e.g., it will no longer be sufficient to consider the first half only of the WAHUHA four-pulse cycle.

As the reader may anticipate compensation schemes have also been devised to fight the effects of this class of pulse imperfections.

Before we continue, let us clarify what exactly we understand by individual flip angle and phase errors, and by phase transients. Our multiple-pulse sequences typically consist of four types of pulses, labeled x, $-x$, y, $-y$. So far we have assumed the flip angle $\beta = \omega_1 t_w$ to be the same for all four types of pulses. We speak of individual flip angle errors when this assumption no longer holds and when the flip angles—which we now denote by β_j, $j = \pm x, \pm y$—differ for the four different types of pulses.

Furthermore, we have assumed tacitly that the rf phases of the four different types of pulses are exactly in quadrature. Of course, designers of multiple-pulse spectrometers—often we experimenters ourselves—cannot guarantee that this is really the case. In our considerations we must therefore allow for deviations γ_j from exact quadrature phases. These we call phase errors of individual pulses. Finally, there are phase transients, which are components of rf pulses that are out of phase with respect to the stationary parts of the pulses. They appear during the rise and fall of rf pulses. Their origin and effects upon the nuclear magnetization $\langle \mathbf{M} \rangle$ are not as obvious as they are for the other two types of pulse imperfections of this class. We feel they should be dealt with in more detail than can be done in a few sentences, and therefore we devote Appendix D to a discussion of phase transients and their effects upon $\langle \mathbf{M} \rangle$.

Our approach to this class of pulse imperfections is to consider the propagation operators $U_p(t)$ corresponding to nonideal pulses, and $U_{rf}(t)$ corresponding to nonideal cycles. $U_{rf}(t)$ will, in general, not be unity for $t = t_c$. We shall represent $U_{rf}(t_c)$ in the form of $\exp[-it_c \mathscr{H}_{err}]$ and shall call $\mathscr{H}_{err}$ the average pulse error Hamiltonian. $\mathscr{H}_{err}$ contains the direct effect of the pulse imperfections upon the motion of $\langle \mathbf{M} \rangle$. There are also indirect effects: As $U_{rf}(t)$ is modified with respect to the ideal case, $\mathscr{H}^{int}_{secular}(t)$, $\mathscr{H}_{int}$, and $\mathscr{H}^{(n)}_{int}$ are also modified (see Section D,2,e).

a. Propagation Operator for Single Nonideal Pulse

The propagation operator for an ideal 90° x pulse is $\exp[\frac{1}{2}i\pi I_x]$, which represents a rotation through $\frac{1}{2}\pi$ about the rotating frame x axis. The propagation operator for a nonideal 90° x pulse that, however, has no phase error can be represented by $e^{i\alpha_{tr}I_y}e^{i\epsilon_x I_x}e^{\frac{1}{2}i\pi I_x}e^{i\alpha_l I_y}$. This represents

a rotation through α_l about the y axis, followed by
a rotation through $\frac{1}{2}\pi + \varepsilon_x$ about the x axis, and
a rotation through α_{tr} about the y axis.

(See Fig. 5-8.) We assume that $B_x(t)$ and $B_y(t)$ do not overlap. This assumption is justified by the very intention of our considerations: We inquire into the nature of small but disturbing effects and look for means to minimize them, but we are not striving for quantitative comparisons between theory and experiment.

$$\alpha_l = \gamma_n \int_l B_y(t)\,dt \qquad (l,\text{ leading edge}),$$

and

$$\alpha_{tr} = \gamma_n \int_{tr} B_y(t)\,dt \qquad (tr,\text{ trailing edge}),$$

arise from phase transients, and

$$\varepsilon_x = \gamma_n \int_{\text{pulse}} B_x(t)\,dt - \tfrac{1}{2}\pi$$

is the flip angle error of the pulse.

A phase error by an angle γ_x is taken into account by sandwiching the above sequence of exponentials between $e^{-i\gamma_x I_z}$ and $e^{+i\gamma_x I_z}$, so that finally the propagation operator $U_p{}^x$ of a nonideal 90° x pulse, including a possible phase error, is expressed by

$$U_p{}^x = e^{-i\gamma_x I_z} e^{i\alpha_{tr} I_y} e^{i\varepsilon_x I_x} e^{\frac{1}{2}i\pi I_x} e^{i\alpha_l I_y} e^{i\gamma_x I_z}. \tag{5-11}$$

b. *Propagation Operator for Nonideal WAHUHA Cycle*

This is "simply" obtained by writing the sequence of 24 exponentials corresponding to

$$\begin{aligned}
U_{rf}(t_c) &= U_p^{+x} U_p^{-y} U_p^{+y} U_p^{-x}\\
&= e^{-i\gamma_x I_z} e^{i\alpha_{tr} I_y} e^{i\varepsilon_x I_x} e^{\frac{1}{2}i\pi I_x} e^{i\alpha_l I_y} e^{i\gamma_x I_z}\\
&\quad \times e^{-i\gamma - y I_z} e^{i\alpha_{tr} I_x} e^{-i\varepsilon - y I_y} e^{-\frac{1}{2}i\pi I_y} e^{i\alpha_l I_x} e^{i\gamma - y I_z}\\
&\quad \times e^{-i\gamma_y I_z} e^{-i\alpha_{tr} I_x} e^{i\varepsilon_y I_y} e^{\frac{1}{2}i\pi I_y} e^{-i\alpha_l I_x} e^{i\gamma_y I_z}\\
&\quad \times e^{-i\gamma - x I_z} e^{-i\alpha_{tr} I_y} e^{-i\varepsilon - x I_x} e^{-\frac{1}{2}i\pi I_x} e^{-i\alpha_l I_y} e^{i\gamma - x I_z}. \tag{5-12}
\end{aligned}$$

We have assumed that the phase transients are identical relative to the main pulses for all four types of pulses (see, however, Appendix D).

Equation (5-12) looks frightening but is not, provided all pulse imperfections are small as we shall assume henceforth. To handle Eq. (5-12) we first shift the factor $e^{-\frac{1}{2}i\pi I_y}$ to the right until it meets and cancels $e^{\frac{1}{2}i\pi I_y}$. Whenever, on its way to the right, it passes spin operators I_x and I_z it transforms

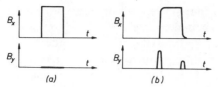

FIG. 5-8. (a) Ideal and (b) nonideal P_x^{90} pulse. The rf field in the rotating frame, $\mathbf{B}_R^{rf}(t) = [B_x(t), B_y(t), 0]$ is shown.

them into $-I_z$ and I_x, respectively.[98] Then we shift the factor $e^{\frac{1}{2}i\pi I_x}$ to the right until it meets and cancels $e^{-\frac{1}{2}i\pi I_x}$. Each I_y passed it transformed into $-I_z$, each I_z into I_y; I_x, of course, remains unaffected. The remaining 20 exponentials all represent small rotations, which we can simply combine in one single exponent, obtaining

$$
\begin{aligned}
U_{rf}(t_c) = \exp\{ & iI_x[(\varepsilon_x - \varepsilon_{-x}) + (\gamma_{-y} - \gamma_y) + (\alpha_{tr} - \alpha_l)] \\
& + iI_y[(\gamma_x - \gamma_{-x} + \gamma_y - \gamma_{-y}) + 2(\alpha_{tr} - \alpha_l)] \\
& + iI_z[(\varepsilon_{-y} - \varepsilon_y) + (\gamma_{-x} - \gamma_x) + (\alpha_{tr} - \alpha_l)]\}.
\end{aligned}
$$
(5-13)

Expressing $U_{rf}(t_c)$ as $\exp[-it_c \mathscr{H}_{err}]$ implies

$$
\begin{aligned}
\mathscr{H}_{err} = -\frac{1}{t_c}\{ & I_x[(\varepsilon_x - \varepsilon_{-x}) - (\gamma_y - \gamma_{-y}) + (\alpha_{tr} - \alpha_l)] \\
& + I_y[(\gamma_x - \gamma_{-x}) + (\gamma_y - \gamma_{-y}) + 2(\alpha_{tr} - \alpha_l)] \\
& + I_z[(\varepsilon_{-y} - \varepsilon_y) - (\gamma_x - \gamma_{-x}) + (\alpha_{tr} - \alpha_l)]\}.
\end{aligned}
$$
(5-14)

By shifting and canceling the factors $\exp[\pm\frac{1}{2}i\pi I_x]$ and $\exp[\pm\frac{1}{2}i\pi I_y]$, and by collecting the exponents of the remaining exponentials in one single exponent, we save the concept of cycles in the presence of errors of individual pulses. Whereas the first step is rigorous, the second is not. It is justified only when all pulse errors are small.

Equation (5-14) expresses a rotation $\boldsymbol{\omega}$ that adds—vectorially—to the rotation $\boldsymbol{\Omega}$ imposed on $\langle\mathbf{M}\rangle$ by $\mathscr{H}_{off}$. $\boldsymbol{\omega}$ can be decomposed into components $\boldsymbol{\omega}_\parallel$ and $\boldsymbol{\omega}_\perp$, which are, respectively, parallel and perpendicular to $\boldsymbol{\Omega}$:

$$
\boldsymbol{\omega}_\parallel = \frac{1}{3}[(\varepsilon_x - \varepsilon_{-x}) - (\varepsilon_y - \varepsilon_{-y}) + 4(\alpha_{tr} - \alpha_l)]\begin{bmatrix} 1 \\ 1 \\ 1 \end{bmatrix},
$$
(5-15)

$$
\boldsymbol{\omega}_\perp = \frac{1}{3}\begin{bmatrix} 2(\varepsilon_x - \varepsilon_{-x}) + (\varepsilon_y - \varepsilon_{-y}) - 3(\gamma_y - \gamma_{-y}) - (\alpha_{tr} - \alpha_l) \\ -(\varepsilon_x - \varepsilon_{-x}) + (\varepsilon_y - \varepsilon_{-y}) + 3(\gamma_x - \gamma_{-x}) + 3(\gamma_y - \gamma_{-y}) + (\alpha_{tr} - \alpha_l) \\ -(\varepsilon_x - \varepsilon_{-x}) - 2(\varepsilon_y - \varepsilon_{-y}) - 3(\gamma_x - \gamma_{-x}) - (\alpha_{tr} - \alpha_l) \end{bmatrix}.
$$
(5-16)

[98] Recall that not only, e.g., $e^{-\frac{1}{2}i\pi I_y} I_z e^{\frac{1}{2}i\pi I_y} = I_x$, but also $e^{-\frac{1}{2}i\pi I_y} e^{i\gamma I_z} e^{\frac{1}{2}i\pi I_y} = e^{i\gamma I_x}$.

Note that neither $\omega_{\parallel}$ nor $\omega_{\perp}$ depends on one type of pulse errors only. This is in contrast to the phase-alternated $(P_x^{90} - P_{-x}^{90})_n$-cycle for which $\omega_{\parallel}$ depends on $(\alpha_{tr} - \alpha_l)$ only (see Appendix D). The phase-alternated cycle may therefore be used to demonstrate the existence and consequences of phase transients.[99] It may also be used to monitor sensitively the symmetry of phase transients $(\alpha_{tr} = \alpha_l)$.

For the WAHUHA cycle the effects of $\omega_{\parallel}$ and $\omega_{\perp}$ on the NMR signal can best be described with reference to Fig. 5-9. $\omega_{\parallel}$ simply adds as a constant to Ω, which itself is proportional to the resonance offset $\Delta\omega$. $\omega_{\perp}$ causes a deviation from the linear relationship between $\Delta\omega_{\text{multiple pulse}}$ and $\Delta\omega$. This excludes practically the central part of the spectrum around $\Delta\omega = 0$ for spectroscopic purposes. In the same region of the spectrum there are—almost unpredictable—variations of the NMR signal amplitude, since near $\Delta\omega = 0$ the direction of $\boldsymbol{\omega} + \boldsymbol{\Omega}$ varies strongly with $\Delta\omega$. We have seen in Section D,1,b that the central part of the spectrum is potentially the most favorable, since the direct line broadening due to the rf inhomogeneity vanishes for $\Delta\omega = 0$ and stays very small when $\Delta\omega$ is small. Just how large the unexploitable part of the spectrum is depends, of course, on the size of $\omega_{\perp}$. We therefore strive always to minimize $\omega_{\perp}$ (and also, of course, $\omega_{\parallel}$). One way of eliminating the harmful effects of phase transients is described in Appendix D. The influence of $\omega_{\perp}$ on $\Delta\omega_{\text{multiple pulse}}$ vanishes for $\Delta\omega \gg |\omega_{\perp}|$. This is a simple consequence of vector addition, but it may also be considered as a manifestation of off-resonance averaging.

c. *Average Pulse Error Hamiltonian for MREV Sequence*

This is one-half the sum of the average pulse error Hamiltonians from each of the two subcycles of the MREV cycle. For the first subcycle, Eq. (5-14) is,

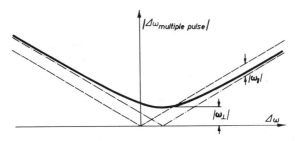

FIG. 5-9. Beat frequency $|\Delta\omega_{\text{multiple pulse}}|$ of the NMR signal in a WAHUHA experiment versus resonance offset $\Delta\omega$. No pulse errors, $---$ (slope $1/\sqrt{3}$); $\omega_{\parallel} \neq 0$, $\omega_{\perp} = 0$, $-\cdot-$; $\omega_{\parallel} \neq 0$, $\omega_{\perp} \neq 0$, ———.

[99] W.-K. Rhim, D. D. Elleman, L. B. Schreiber, and R. W. Vaughan, *J. Chem. Phys.* **60**, 4595 (1974).

of course, obtained again, and with some reversals of signs, Eq. (5-14) is also obtained for the second subcycle. In total, one gets

$$\mathcal{H}_{err}(MREV) = -\frac{2}{t_c}\{I_x[(\varepsilon_x - \varepsilon_{-x}) - (\gamma_y - \gamma_{-y}) + (\alpha_{tr} - \alpha_l)]$$

$$+ I_y[\gamma_x - \gamma_{-x}] + I_z[\alpha_{tr} - \alpha_l]\}. \tag{5-17}$$

Equation (5-17) tells us that the MREV cycle causes some cancellation of pulse errors; however, none of the pulse errors we are discussing currently is compensated fully. In the following subsection we shall turn to a technique by which it is possible to compensate for both flip angle and phase errors of individual pulses. It works for both the WAHUHA and the MREV sequences and is certainly even more general.

d. *Flip Angle and Phase Compensation*[55, 100]

The idea of the technique is to exchange the misalignments γ_x, ε_x and γ_{-x}, ε_{-x} on the one hand, and γ_y, ε_y and γ_{-y}, ε_{-y} on the other hand, in alternating cycles. This leads to compensation as $\mathcal{H}_{err}$ depends only on the differences $(\gamma_j - \gamma_{-j})$ and $(\varepsilon_j - \varepsilon_{-j})$, $j = x, y$. In order to understand how the technique works, we must briefly consider how quadrature phase rf pulses are typically generated in multiple-pulse spectrometers (Fig. 5-10).

One starts with a crystal oscillator or something equivalent, e.g., a frequency synthesizer; this drives a four-port power splitter the output of which drives, in turn, four identical channels each consisting of a phase shifter Φ_{j*} and an rf gate G_{j*}. The rf gates are activated by logic pulses. The rf pulses generated in the four channels are subsequently combined and then amplified in a single-channel power amplifier chain.

To achieve phase compensation a phase switcher is inserted between the

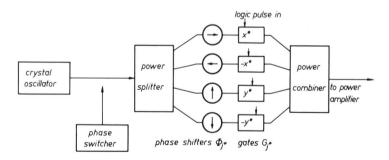

FIG. 5-10. Generation of quadrature phase rf pulses.

[100] Only phase compensation has been described in the literature, while it seems to have passed unnoticed that the same technique also compensates for flip angle errors of individual pulses.

oscillator and the power splitter. Technically the phase switcher consists of nothing more than a double balanced mixer whose X port is driven with either positive or negative current. It is also activated by logic pulses. When it is switched from "off" to "on" it reverses the phase of the input signal of the power splitter and thus the phase of any pulse generated in any of the four channels.

We have labeled the four channels x^*, $-x^*$, y^*, and $-y^*$. We shall continue to label the pulses that the nuclear spins see by x, $-x$, y, and $-y$. So far in this text there has been no need for this distinction.

By definition, the x^*, $-x^*$, y^*, $-y^*$ channels generate x, $-x$, y, $-y$ pulses, respectively, when the phase switcher is in its off state. Provided the phase switcher is ideal, the x^*, $-x^*$, y^*, $-y^*$ channels generate $-x$, x, $-y$, y pulses when the phase switcher is in its on state.

Now consider Fig. 5-11, which shows an extended cycle of a phase-compensated WAHUHA sequence. The top row shows the timing of the logic pulses that activate the rf gates. The first four pulses clearly generate a WAHUHA four-pulse cycle. The second quartet of pulses activates the x^*, $-x^*$, and the y^*, $-y^*$ gates, respectively, in reversed order. Since, however, the phase switcher is on during the same time (see row 2) the nuclear spins see just another WAHUHA four-pulse cycle (row 3). Row 4 shows schematically the pulses that the nuclear spins see for a case where the phase shifters are exactly in quadrature, and where the phase switcher works perfectly. We hope

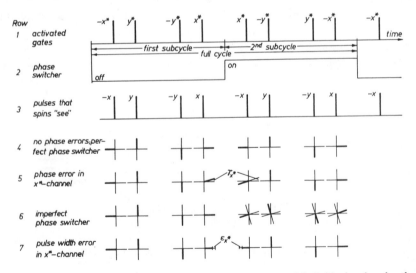

FIG. 5-11. Compensation of pulse width and phase errors of individual pulses by the phase switching technique. Rows 4–7 are views of the xy plane of the rotating frame. Pulses are shown as heavy lines.

that the symbols in the diagram are self-explanatory. In row 5 we have depicted a case where the phase in the x^* channel is misaligned but everything else is still perfect. Note, in particular, that this misalignment affects the x pulse of the first, but the $-x$ pulse of the second subcycle.

We are now ready to explain phase compensation; we do it with the following chain of arguments:

(i) The difference $\gamma_x - \gamma_{-x}$ has the same magnitude but the opposite sign for the first and second four-pulse subcycles of the full eight-pulse cycle.

(ii) $\mathscr{H}_{err}$(full cycle) $= \frac{1}{2}[\mathscr{H}_{err}$(first subcycle) $+ \mathscr{H}_{err}$(second subcycle)].

(iii) $\mathscr{H}_{err}$ of any subcycle depends on γ_x and γ_{-x} only in the form of $(\gamma_x - \gamma_{-x})$ [see Eq. (5-14)].

It follows from (i)–(iii) that a phase misalignment in the x channel is compensated over the full cycle:

(iv) Nothing is special with the x channel; phase misalignments are compensated therefore in all channels.

(v) The linearity of $\mathscr{H}_{err}$ with respect to all γ_j [cf. Eqs. (5-14) and (5-17)] implies that simultaneous phase misalignments in all channels are compensated. Recall, however, that the linearity invoked is an approximation valid only for small misalignments.

Row 6 in Fig. 5-11 shows that failure of the phase switcher to switch the rf phase by exactly 180° is not harmful since the differences $\gamma_j - \gamma_{-j}$ in each subcycle are not affected.

Row 7 shows a case where the flip angle in the x^* channel is misaligned. The consequence is that in the first subcycle it is ε_x that is nonzero, whereas it is ε_{-x} that is nonzero in the second subcycle. A completely analogous chain of arguments as before leads to the conclusion that the phase-switching technique in alternating cycles also leads to a compensation of flip angle errors of individual pulses.

We have implemented the phase-switching technique on our spectrometer and our experiments definitely confirm the above conclusions. We do find that the behavior of the NMR signal becomes much less critically dependent on the settings of all pulse width and phase controls when we switch from the ordinary to the compensated versions of either the WAHUHA or MREV sequences. All of our recent work is done with phase and flip angle compensated cycles.

So far we have discussed only the direct effects of flip angle and phase errors of individual pulses, and of phase transients. Although these effects are very disturbing unless they are carefully minimized and compensated for, they do not affect the spectral resolution. As mentioned already there are also indirect effects that do affect the spectral resolution. We turn to them now.

e. Line Broadening Resulting from Errors of Individual Pulses

Pulse errors affect the propagation operator $U_{rf}(t)$ and hence indirectly the residual dipolar line broadening and the scaling factor. As $U_{rf}(t)$ does not return to unity for $t = t_c$ in the presence of pulse errors different for different pulses, we must be cautious in applying the average Hamiltonian theory. Therefore, we go back one step in the theory and consider the full propagation operator $U_R(t)$ of the density matrix $\rho_R(t)$ in the rotating frame. It can be written directly, and particularly easily when $t_W \ll \tau$ as we shall assume from now on. For $t = t_c$ it is

$$U_R(t_c) = \exp(-i\mathscr{H}_R^{int}\tau)\, U_p^{\,x}\exp(-i\mathscr{H}_R^{int}\tau)\, U_p^{-y}\exp(-i\mathscr{H}_R^{int}2\cdot\tau)$$
$$\times\ U_p^{\,y}\exp(-i\mathscr{H}_R^{int}\tau)\, U_p^{-x}\exp(-i\mathscr{H}_R^{int}\tau) \qquad (5\text{-}18)$$

where $U_p^{\,x}$, e.g., is given by Eq. (5-11). $U_R(t_c)$ consists of exponentials with three types of exponents:

(i) *Large* exponents: $\frac{1}{2}i\pi I_x,\ -\frac{1}{2}i\pi I_x,\ \frac{1}{2}i\pi I_y,\ -\frac{1}{2}i\pi I_y$.

Under usual conditions the exponents of all other exponentials are small. There are two types of small exponents:

(ii) $-i\mathscr{H}_R^{int}\tau$, and
(iii) exponents containing pulse error angles $(\pm\gamma_j,\varepsilon_j,\alpha_l,\alpha_{tr})$.

These originate, of course, also in Hamiltonians (pulse error Hamiltonians[101] $\mathscr{H}_\gamma,\mathscr{H}_\varepsilon,\mathscr{H}_\alpha$) multiplied by or integrated over some (short) time interval. These Hamiltonians distinguish themselves from $\mathscr{H}_R^{int}$ in one respect: They are time dependent, they act upon the spin system only intermittently, whereas $\mathscr{H}_R^{int}$ is time independent. This difference disappears after shifting, as before, the exponentials with large exponents until they meet and cancel. What they leave behind in the exponentials with small exponents is $\tilde{\mathscr{H}}(t)$, which consists of $\mathscr{H}_R^{int}(t)$ and $\mathscr{H}_{err}^{rf}(t)$, both having equal roles and rights.

Of course we wish to express $U_R(t_c)$ in the form of a single exponential. We obtain the exponent with the aid of the Magnus expansion (cf. Appendix B). The zeroth-order term $\bar{\mathscr{H}}$ consists of $\bar{\mathscr{H}}_{int}$ and $\bar{\mathscr{H}}_{err}$. Both have been discussed in detail already. We are interested here in the first-order correction term $\bar{\mathscr{H}}^{(1)}$, in fact, only in the first-order cross term between $\mathscr{H}_R^{int}(t)$ and $\mathscr{H}_{err}^{rf}(t)$, which contains the lowest order line broadening effects resulting from errors of individual pulses.

Clearly, we can evaluate the cross term between each of the internal Hamiltonians with each of the pulse error Hamiltonians separately. We indicate the

---

[101] The flip angle error Hamiltonian $\mathscr{H}_\varepsilon$ may be considered to act while the pulses are on, the phase transient Hamiltonian $\mathscr{H}_\alpha$ during rise and fall of the pulses, while in our description of the phase errors $\mathscr{H}_\gamma$ acts only during infinitely short intervals before and after the pulses.

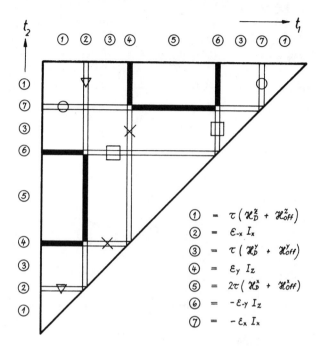

FIG. 5-12. $\mathscr{H}^{(1)}$ cross terms between $\mathscr{H}_D + \mathscr{H}_{off}$ and $\mathscr{H}_\varepsilon$ for WAHUHA cycle:

$$\mathscr{H}^{(1)} = -\frac{i}{2t_c}\int_0^{t_c}dt_2\int_0^{t_2}dt_1[\mathscr{H}(t_2),\mathscr{H}(t_1)]$$

$$\rightarrow -\frac{i}{2t_c}\{2[\textcircled{3},(\textcircled{2}-\textcircled{7})]+[\textcircled{5},(\textcircled{4}-\textcircled{6})]\}.$$

Note: Only the narrow strips are candidates for contributions. Dark strips give no contributions. Contributions from strips marked by O, ×, ▽, □ cancel.

procedure for the cross terms between $\mathscr{H}_D$ and $\mathscr{H}_{off}$ on the one hand, and the flip angle Hamiltonian $\mathscr{H}_\varepsilon$ on the other hand. We denote them by $\mathscr{H}_{D,\varepsilon}^{(1)}$ and $\mathscr{H}_{off,\varepsilon}^{(1)}$. Figure 5-12 is helpful for sorting out the relevant terms. By mere inspection we find for the WAHUHA sequence

$$\mathscr{H}_{D,\varepsilon}^{(1)} + \mathscr{H}_{off,\varepsilon}^{(1)} = -\frac{i}{2t_c}\{2[\textcircled{3},(\textcircled{2}-\textcircled{7})]+[\textcircled{5},(\textcircled{4}-\textcircled{6})]\} \tag{5-19}$$

$$= (-i/12\tau)\{2[(\mathscr{H}_D^y+\mathscr{H}_{off}^y),I_x]\tau(\varepsilon_x+\varepsilon_{-x}) + [(\mathscr{H}_D^x+\mathscr{H}_{off}^x),I_z]2\tau(\varepsilon_y+\varepsilon_{-y})\},$$

$$\mathscr{H}_{off,\varepsilon}^{(1)} = -\tfrac{1}{6}\Delta\omega\{(\varepsilon_x+\varepsilon_{-x})I_z+(\varepsilon_y+\varepsilon_{-y})I_y\}, \tag{5-20 WAHUHA}$$

$$\mathcal{H}_{D,\varepsilon}^{(1)} = \tfrac{1}{2} \sum_{i<k} b^{ik} \{ (\varepsilon_x + \varepsilon_{-x})(I_y{}^i I_z{}^k + I_z{}^i I_y{}^k) + (\varepsilon_y + \varepsilon_{-y})(I_x{}^i I_z{}^k + I_z{}^i I_x{}^k) \}.$$

(5-21 WAHUHA)

We leave it as an exercise for the reader to convince himself that the cross terms between $\mathcal{H}_D$ and $\mathcal{H}_{\text{off}}$ on the one hand, and the phase error and phase transient Hamiltonians on the other hand, are given by

$$\mathcal{H}_{\text{off},\gamma}^{(1)} = \tfrac{1}{6} \Delta\omega \{ (\gamma_x + \gamma_{-x})(I_y - I_x) + (\gamma_x + \gamma_{-x} - \gamma_y - \gamma_{-y})I_z \},$$

(5-22 WAHUHA)

$$\mathcal{H}_{D,\gamma}^{(1)} = \tfrac{1}{2}(\gamma_y + \gamma_{-y} - \gamma_x - \gamma_{-x}) \sum_{i<k} b^{ik} (I_z{}^i I_x{}^k + I_x{}^i I_z{}^k),$$

(5-23 WAHUHA)

$$\mathcal{H}_{\text{off},\alpha}^{(1)} = \tfrac{1}{6} \Delta\omega (\alpha_l + \alpha_{\text{tr}})(I_x - I_y),$$ (5-24 WAHUHA)

$$\mathcal{H}_{D,\alpha}^{(1)} = 0.$$ (5-25 WAHUHA)

Equations (5-20) and (5-21) also cover the linear effects, but only the linear effects of the multiple-pulse line broadening resulting from flip angle errors, which are common to all pulses. We have discussed such errors in detail in Section D,1,b, not restricting ourselves to terms linear in ε. There they appeared in the average Hamiltonian itself (in contradistinction to $\mathcal{H}^{(1)}$) because common flip angle errors could be included in $U_{\text{rf}}(t)$ since they do not destroy the cyclic property of multiple-pulse sequences. We now recognize that it is the average of ε_x, ε_{-x} and ε_y, ε_{-y} that determines the size of the effects. In Section D,1,b we saw that the coefficients of those terms in $\mathcal{H}_D$ that are bilinear in the spin operations and linear in ε drop out with the MREV sequence. Analogously it is found here that

$$\mathcal{H}_{D,\varepsilon}^{(1)}(\text{MREV}) = 0.$$ (5-26 MREV)

$\mathcal{H}_{D,\gamma}^{(1)}$ expresses the residual dipolar line broadening in WAHUHA experiments that results from a failure to adjust the rf phases of the $+x$, $-x$, $+y$, and $-y$ pulses exactly in quadrature. We recognize that the rf fields of the $\pm x$ pulses on the one hand, and those of the $\pm y$ pulses on the other hand, must be perpendicular *on the average* for optimum resolution.

This suggests—together with other, more practical reasons—that it is advisable to design the "quadrature phase box" in multiple-pulse spectrometers in such a way as to enable the experimenter to shift the rf phases of the $+x$ and $-x$ pulses simultaneously relative to the phases of the $+y$ and $-y$ pulses, and vice versa.

$\mathcal{H}_{D,\gamma}^{(1)}$ vanishes for the MREV eight-pulse cycle. This is no surprise since the roles of the $+x$ and $-x$ pulses—which determine the signs of the γ_j in Eq. (5-23)—are opposite in the first and second four-pulse subcycles of the full MREV eight-pulse cycle.

It is both interesting and reassuring to note that phase transients do not entail first-order residual dipolar line broadening [see Eq. (5-25)] in WAHUHA four-pulse, but of course also in MREV eight-pulse experiments. Whereas only the difference of the leading and trailing edge phase transients, $\alpha_l - \alpha_{tr}$, enters the expressions for the average phase transient Hamiltonians [see Eqs. (5-14) and (5-17)], it is the sum $\alpha_l + \alpha_{tr}$ that enters $\mathcal{H}_{off,\alpha}^{(1)}$. At first sight this looks rather disturbing, particularly for narrow-band systems, where the phase transients constitute a substantial fraction of the main pulse itself. However, it only looks disturbing while, in fact, it is not because of two reasons:

(i) $\mathcal{H}_{off,\alpha}^{(1)}$ contains $\Delta\omega$ as a factor; and
(ii) $(I_x - I_y)$ is "perpendicular" to $(I_x + I_y + I_z)$.

When $\Delta\omega$ is small $\mathcal{H}_{off,\alpha}^{(1)}$ is also small, and when $\Delta\omega$ is large, off-resonance averaging will average out $\mathcal{H}_{off,\alpha}^{(1)}$.

For the MREV sequence,

$$\mathcal{H}_{off,\alpha}^{(1)} = \tfrac{1}{6}\,\Delta\omega\{3(\alpha_l - \alpha_{tr})(I_x - I_z) - (\alpha_l + \alpha_{tr})I_y\}. \qquad \text{(5-27 MREV)}$$

Note that both $(I_x - I_z)$ and I_y are "perpendicular" to $I_x + I_z$, which gives the direction of the resonance offset rotation axis for the MREV sequence. Off-resonance averaging will therefore average out $\mathcal{H}_{off,\alpha}^{(1)}(\text{MREV})$.

It may be surprising that $\mathcal{H}_{off,\alpha}^{(1)}(\text{MREV})$ depends on both $\alpha_l + \alpha_{tr}$ and $\alpha_l - \alpha_{tr}$, whereas $\mathcal{H}_{off,\alpha}^{(1)}(\text{WAHUHA})$ depends only on $\alpha_l + \alpha_{tr}$. The formal reason for the difference is that the average of $\mathcal{H}_R^{off}(t)$ does not vanish. MREV cross terms involving $\mathcal{H}_R^{off}(t)$ are, for the same reason, in general not simply the sums of the corresponding WAHUHA terms with some reversals of signs properly taken into account.

With these remarks we close our discussion of pulse errors and turn briefly to timing errors.

3. TIMING ERRORS

So far we have assumed tacitly that the timing of the pulses in both the WAHUHA and MREV sequences is perfect. What happens if it is not? Again we first consider a WAHUHA sequence. We suppose that the $-x$ pulse, for example, arrives late by $\delta\tau$ (see Fig. 5-13). It is immediately clear that

$$\mathcal{H}_D = (\delta\tau/t_c)(\mathcal{H}_D^z - \mathcal{H}_D^y) \neq 0. \qquad (5\text{-}28)$$

Now let us ask how small must we keep $\delta\tau$ for the residual dipolar line broadening resulting from timing errors to remain negligible. Dipolar line broadening in solids can be reduced by multiple-pulse sequences to below 1%— this can be achieved now quite routinely—but we would like to do better by at least one order of magnitude.

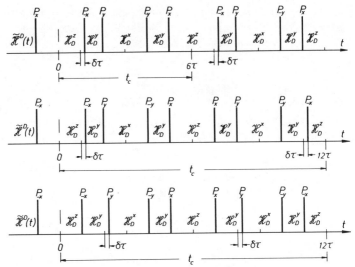

FIG. 5-13. Timing errors. (*top row*) WAHUHA sequence with the $-x$ pulse coming late by $\delta\tau$. (*center row*) MREV sequence with the same timing error. $\mathscr{H}_\mathrm{D}$ is not changed. (*bottom row*) MREV sequence with the $+y$ pulse coming late by $\delta\tau$. $\mathscr{H}_\mathrm{D}$ is changed.

Consequently, $\delta\tau/t_\mathrm{c}$ should be kept smaller than 10^{-3}. For $t_\mathrm{c} = 20$ μsec this means that $\delta\tau$ must be kept smaller than 20 nsec.

In modern spectrometers the timing of the pulses is usually derived from a crystal oscillator (master clock); therefore, the accuracy of the timing seems to be more than adequate. However, the $+x$, $-x$, $+y$, $-y$ pulses are typically generated in different channels consisting of a lot of logic and rf circuitry. All gates, flip–flops, monoflops, etc., inherently have delays that are of the order of 10–30 nsec for the currently preferred TTL logic chips. In our spectrometer these delays add up to more than 100 nsec. They would not cause any harm if they were equal in all channels. However, they cannot be expected to be precisely equal since the delays are subject to variations from chip to chip. The differences of the delays in the different channels may well reach the order of magnitude of 20 nsec unless special care is taken of that problem in the design of the spectrometer.

Pulse-timing errors are, however, not as critical as they look. Off-resonance averaging comes to our rescue once again. The form of Eq. (5-28) implies that $\mathscr{H}_\mathrm{D}$ is averaged out by WAHUHA off-resonance averaging, provided $\Delta\omega$ is large enough.

Things are a little more involved with the MREV sequence: Timing errors of the $+x$ and $-x$ pulses are compensated over the full cycle and are thus not critical (see Fig. 5-13, center row). Timing errors of the $+y$ and $-y$ pulses are not compensated (Fig. 5-13, bottom row). If the y pulses, e.g., arrive late by $\delta\tau$ the average dipolar Hamiltonian becomes

$$\mathscr{H}_{\mathrm{D}}(\mathrm{MREV}) = 2(\delta\tau/t_{\mathrm{c}})(\mathscr{H}_{\mathrm{D}}{}^{y} - \mathscr{H}_{\mathrm{D}}{}^{x}). \qquad (5\text{-}29)$$

The origin of the difference between the $\pm x$ and $\pm y$ pulses with respect to their susceptibility to timing errors is that in the MREV sequence two $+x$ and two $-x$ pulses are always fired in succession, whereas the $+y$ and $-y$ pulses are always followed by a different kind of pulse.

MREV off-resonance averaging reduces (by roughly a factor of $\frac{1}{2}$) but does not suppress completely the residual dipolar line broadening resulting from timing errors. As was the case with the second-order dipolar correction term $\mathscr{H}_{\mathrm{D}}^{(2)}$ (see Section B) MREV off-resonance averaging is not as effective as is WAHUHA off-resonance averaging in removing residual dipolar line broadening resulting from timing errors.

Finally, we would like to remark that the flip angle—and phase—compensation scheme described above—if designed properly!—is also able to compensate for pulse-timing errors. The reader may convince himself of this fact by combining Figs. 5-11 and 5-13.

E. What Is the Resolution-Limiting Factor in Multiple-Pulse Experiments?

We have dealt with eight different imperfections of multiple-pulse trains, which all could be suspected to affect adversely the resolution of solid state multiple-pulse spectra. With one exception—line broadening arising from rf inhomogeneity mediated by the flip angle dependence of the scaling factor—we always arrived at the conclusion that the troublesome effects can be overcome or "compensated" by one trick or the other.

Indeed, we have experimentally reached a state where to the best of our knowledge the experimental results no longer depend noticeably on pulse imperfections or misalignments. This statement can be checked by artificially increasing the various known pulse imperfections.

Thus we find that the spectral resolution does not change to a degree worth mentioning if we vary the ratio t_{W}/τ over a range from, say, 0.15 to 0.30—provided the proper flip angle β_0 is always chosen.

When using a flip angle-compensated cycle, e.g., the MREV eight-pulse cycle, we also find that the results are quite insensitive to small flip angle errors. The test is to vary slightly the transmitter power and to watch what happens.

Similarly, when using a flip angle-compensated, phase-compensated cycle, we find that the multiple-pulse response of solid samples remains virtually unaffected when we change slightly any one of the eight individual phase or pulse width controls.

The same is true of phase transients: While the on-resonance or close-to-resonance beat structure of the multiple-pulse signal is very sensitive to nonsymmetric components of the leading and trailing edge phase transients, the line widths of off-resonance lines are not.

Finally, we have artificially introduced a pulse-timing error into a WAHUHA sequence by simply inserting a piece of rf cable, corresponding to a pulse delay of about 30 nsec, into the $-x$ pulse line. Judging from the multiple-pulse oscilloscope traces we could not detect any degradation of the results.

Thus we conclude that multiple-pulse trains can be controlled to a degree where pulse imperfections are no longer responsible for most of the line broadening observed in multiple-pulse spectra.

So, what then limits the resolution? Unfortunately there seems to be no simple general answer to this question. (Hopefully this statement is not a mere self-flattering rephrasing of "I do not know.") We shall cite now some specific experiments where the factors that limit the resolution appear to have been tracked down.

1. CaF_2, PROTOTYPE OF COMPOUNDS THAT CONTAIN ONE KIND OF SPINS ONLY

CaF_2 is the solid state NMR prototype of compounds that contain abundant I spins with a large magnetogyric ratio (^{19}F) and essentially no other nuclei with spin quantum number $\neq 0$. For most—not all—purposes the presence of the magnetic ^{43}Ca isotopes (natural abundance 0.13%) can be neglected. A ^{19}F shielding anisotropy does not exist in CaF_2 as the ^{19}F site symmetry is cubic. Much of what we shall say about CaF_2 applies also to other compounds that contain essentially only one kind of spins. To this class belong, in particular, all compounds composed of oxygen, sulfur, carbon, calcium, and either fluorine or hydrogen.

Rhim et al.[89] made a careful study of the multiple-pulse linewidth, or alternatively, of the multiple-pulse decay time of CaF_2 using both the WAHUHA four-pulse and the MREV eight-pulse cycles. Their findings concerning the resolution limiting factors are as follows:

(a) Unless ellipsoidally, in practice spherically, shaped samples are used, the warping of the applied field $\mathbf{B}_{st}$ due to the nonzero susceptibility of the sample limits the resolution. In one particular experiment using the MREV eight-pulse cycle they obtained the following results:

spherical sample (diam = 4 mm), linewidth $\cong$ 0.4 ppm;
cylindrical sample (diam = 4 mm, length = 4.5 mm, axis perpendicular to $\mathbf{B}_{st}$), linewidth $\cong$ 1.1 ppm;

parallelepiped ($4 \times 4 \times 5$ mm, $\mathbf{B}_{st}$ perpendicular to one face), linewidth $\cong 1.4$ ppm.

In high-resolution NMR in liquids, ellipsoids are usually approximated by long cylinders extending well beyond the NMR receiving coil. This is not possible in high-resolution NMR in solids, where for rf power considerations single-coil probe configurations are usually preferred. Samples extending beyond the rf coil would imply excessively large rf inhomogeneities over the sample volume. Thus one is essentially restricted to spheres. Spheres of CaF_2 can easily be made and oriented (!) but shaping spheres of often rather soft single crystals of organic compounds needed for proton work, and subsequently orienting them, can be exceedingly difficult.

Nevertheless, in our recent work[102] on single crystals of ferrocene ($C_5H_5FeC_5H_5$) and *trans*-diiodoethylene ($C_2H_2I_2$) we concluded that for bulk susceptibility reasons we had to get spheres (and we finally succeeded in making them) in order (i) to exploit fully the line-narrowing capability of our spectrometer, and (ii) to measure reliably the proton shift tensors in these compounds (see also Chapter VI, Section C, 2).

(b) Far off resonance the direct line-broadening mechanism arising from the inhomogeneity of the rf field becomes the resolution-limiting factor. The question is, of course, what "far off resonance" means in practice. The answer to this question depends, on the one hand, on the homogeneity of the rf field, and on the other, on the spread of the spectra to be observed.

For samples with single-line spectra such as CaF_2, one can always choose an off-resonance position such that the rf inhomogeneity is not predominant in broadening the line. Working at 60 or 90 MHz the same is usually true of entire proton spectra the spread of which is typically $\leqslant 30$ ppm. However, if one goes up to, say, 300 MHz the situation may become different. The situation is different for ^{19}F spectra even at 60 or 90 MHz, because the typical spread of the spectra is much larger (≈ 150 ppm). At least some of the lines will always be so far off resonance that the resolution may be affected if not limited by the rf inhomogeneity.

(c) In a substantial range of off-resonance frequencies (excluding very small and very large ones) the resolution-limiting factor is τ. (We prefer to characterize the time scale of multiple-pulse sequences by τ rather than t_c, since in contradistinction to the cycle time t_c, τ does not depend on the complexity of the particular pulse sequence.) Rhim and co-workers[89] varied τ from 4.2 to 9.2 μsec and measured the ^{19}F multiple-pulse decay time of CaF_2. For a resonance offset $\Delta\omega = 2\pi \times 0.6$ kHz and $\tau > 5.3$ μsec, the decay time was found to vary as τ^{-4}. When τ was made smaller than 5.3 μsec, the

[102] H. W. Spiess, H. Zimmermann, and U. Haeberlen, *Chem. Phys.*, in press.

decay time no longer varied as rapidly as before, and for $\tau < 4.8$ μsec it remained virtually independent of τ.

When $\Delta\omega$ was increased to $2\pi \times 2.5$ kHz the decay time variation changed to a τ^{-2} dependence and a leveling off was not observed even for the smallest values of τ accessible in this experiment.

These results have been interpreted as follows:

The average Hamiltonian and all odd-order correction terms are zero (see Section A). The first nonvanishing purely dipolar correction term is $\mathscr{H}_D^{(2)}$. It is killed, however, by off-resonance averaging (see Section B). Therefore, the lowest order purely dipolar term that can possibly manifest itself in the decay time is $\mathscr{H}_D^{(4)}$. Its τ-dependence is τ^4, which corresponds to the observed τ^{-4} dependence of the decay time.

The leveling off of the decay time for $\tau < 5.3$ μsec is attributed to other effects independent of τ, such as the inhomogeneity of $\mathbf{B}_{st}$. These effects become competitive when the dipolar line broadening has been suppressed by the multiple-pulse sequence to a very large extent.

On going further off resonance dipolar resonance offset cross terms in $\mathscr{H}^{(2)}$ take over. They increase either linearly or quadratically as $\Delta\omega$ increases and are eliminated only partially by off-resonance averaging. Their τ-dependence is τ^2—as observed by Rhim $et\ al.$ for $\Delta\omega = 2\pi \times 2.5$ kHz. We should add, however, that this explanation of the τ dependence of Rhim's results has been questioned by Garroway $et\ al.$[90] Rhim's experimental results, however, remain unquestionable.

Qualitatively they are confirmed by many other experiments, for example, by our own on malonic acid[103] $CH_2(COOH)_2$. Malonic acid crystallizes in such a way that all molecules are magnetically equivalent. The smallest proton–proton distance is by far the distance between the two methylene protons. Let us call the corresponding internuclear vector $\mathbf{r}_m$. In some orientations of the sample crystals we observed remarkably narrow and well-resolved lines, whereas in others a broad "smear" was all we could get. The explanation is simple; it is contained in the orientation and distance dependence of dipole–dipole interactions: Well-resolved spectra are obtained when $\mathbf{r}_m$ is at or close to the magic angle with respect to $\mathbf{B}_{st}$, and a "smear" is observed when $\mathbf{r}_m$ is parallel or close to parallel to $\mathbf{B}_{st}$.

What is called for is, of course, a maximum exploitation of the information obtainable from the favorable orientations of the crystal. Note that there is a onefold infinity of orientations that all preserve the magic angle between one crystal fixed vector, for example, $\mathbf{r}_m$, and $\mathbf{B}_{st}$.

[103] U. Haeberlen, U. Kohlschütter, J. Kempf, H. W. Spiess, and H. Zimmermann, $Chem.$ $Phys.$ **3**, 248 (1974).

The information obtainable under these optimum conditions is sufficient to determine five out of the six independent parameters of symmetric shielding tensors.[104] Experiments along these lines have been carried out by Haubenreisser and Schnabel[105] on $MgSO_4 \times H_2O$ and by Kohlschütter[104] on malonic acid.

2. $KHSO_4$, Example of Substances Containing Abundant S Spins with Small Magnetogyric Ratio

$KHSO_4$ has been investigated by Pollak-Stachura and the author.[106] The primary purpose of the investigation was the elucidation of the shielding tensors of the two types of protons contained in $KHSO_4$; multiple-pulse linewidths were only a side aspect. Nevertheless, $KHSO_4$ may be considered as a model compound in our present context since it is typical for compounds that contain abundant S spins (potassium) with a small magnetogyric ratio.

All protons in $KHSO_4$ are involved in hydrogen bonds; however, there are two types of hydrogen bonds. In the first type two quasi-parallel hydrogen bonds link two SO_4 tetrahedra which, as a result of this bonding, form dimers. The second type links SO_4 tetrahedra in a chainlike manner. Let us first consider the dimers. The full width at half-height ($\delta\omega$) of the multiple-pulse lines arising from the relevant protons was experimentally found to be about $2\pi \times 360$ Hz. A dependence of the linewidth on the orientation of the crystal was not measurable directly due to heavy overlap of lines; however, an observed variation of about 15% of the height of the lines is indicative of a corresponding variation of the width of the lines.

In an effort to account for the observed linewidths we calculated the contributions to the second moment of the proton lines arising from the K^+ ions. There are eight K^+ ions at a distance from 3.41 to 5.47 Å from the protons in question. Because there are so many K^+ ions at a comparable distance only a slight variation of the second moment with the orientation of the crystal is expected. Therefore, we calculated only a powder average, taking into account individually all K^+ ions at distances <7 Å; for the ions further away we assumed a homogeneous distribution. Assuming, furthermore, a gaussian lineshape and absence of self-decoupling (see Chapter IV, Section F, 5) we arrived at a contribution of the K^+ ions to the proton multiple-pulse linewidth of

$$\delta\omega_K = (2.36/\sqrt{3})\langle\Delta\omega_K^2\rangle^{1/2} = 2\pi \times 320 \text{ Hz}.$$

[104] U. Kohlschütter, Ph.D. Thesis, University of Heidelberg, 1975.
[105] U. Haubenreisser and B. Schnabel, *Proc. 1st Spec. Colloq. AMPERE, 1973* p. 140 (1973).
[106] M. Pollak-Stachura and U. Haeberlen, to be published in *J. Magn. Resonance*.

The factor 2.36 originates in the gaussian lineshape assumption, the factor $1/\sqrt{3}$ in the multiple-pulse scaling of the proton–potassium dipolar interactions. This number compares very favorably with the experimental one. It tells us

(i)   Proton–K^+ dipole–dipole interactions limit the resolution.
(ii) K^+ self-decoupling is ineffective in $KHSO_4$ (which is not surprising).

The (small) dependence of the linewidth on the orientation of the crystal is obviously related to residual proton–proton dipole interactions: The height of the line is clearly at its minimum (the linewidth at its maximum) when the applied field is parallel to the line connecting the two protons of the dimer. Their distance is 2.602 Å. In this orientation of the crystal the strength of the proton–proton interaction reaches its peak value.

The situation is somewhat different for the protons in the chains: There are no close nearby protons; however, there is one potassium ion at a distance of only 2.826 Å. The distance to the next nearest K^+ ion is already 3.440 Å. We again calculated the proton multiple-pulse linewidth contribution due to the K^+ ions. We proceeded as before, with the one difference that we took into account the orientation dependence of the second moment arising from the nearest K^+ ion. For the other K^+ ions we took a powder average as before. The calculation resulted in linewidths $\delta\omega_K$ ranging from $2\pi \times 304$ Hz for the magic-angle direction of the nearest K^+ ion to $2\pi \times 815$ Hz for the most unfavorable case (applied field along the proton/nearest K^+ ion vector, r_{HK}).

At the magic-angle direction of r_{HK} the experiment clearly yielded the narrowest lines ($\delta\omega \approx 2\pi \times 300$ Hz); when r_{HK} was parallel to B_{st} the linewidth increased to about twice its minimum value. Again, these numbers can be only rough estimates due to heavy overlap of lines.

From these observations we conclude:

When the crystal under investigation contains abundant S spins with small magnetogyric ratio such as ^{39}K and ^{41}K, and when there is no S-spin self-decoupling, then it is typically the I–S dipolar coupling that limits the resolution. In comparison with the resolution obtainable in CaF_2, the degradation of resolution can easily be as much as tenfold. Nevertheless, the resolution can still be good enough to measure ^{19}F and even ^1H shielding tensors with reasonable accuracy. Our estimated error limits for the proton shielding tensors (four independent ones) in $KHSO_4$ are about ± 0.5 ppm, while the anisotropy is about 30 ppm. All shielding tensors in $KHSO_4$ are approximately axially symmetric. The error for the unique principal shielding direction is not larger than $\pm 2°$.

F. Further Propositions for Improving the Resolution in Solid State NMR

1. QUASI-CONTINUOUS IRRADIATION

We have seen in the preceding section that τ is often the resolution-limiting factor in multiple-pulse experiments. There are a number of reasons why at present we cannot make τ smaller than about 2 μsec in WAHUHA-type experiments. One is the amount of (stable) rf power needed to flip the nuclear spins through 90° in a very short time interval. Another is the amount of heat dissipated in the probe.

Now, Mehring and Waugh[107] pointed out that the average amount of rf power, P_{av}, to be fed into a given NMR probe for fixed τ decreases on increasing the pulse width t_W. Not realizing in 1972 that coherent averaging also takes place in WAHUHA-type experiments while the pulses are "on" they chose the six-pulse sequence[51] shown in Fig. 5-14 as their starting point.

For fixed τ the average rf power needed for this sequence is directly proportional to t_W^{-1}:

$$P_{av} = P_{peak}\frac{t_W}{\tau}, \qquad P_{peak} \propto B_1^{\ 2} \propto t_W^{-2};$$

hence $P_{av} \propto t_W^{-1}$. The minimum amount of rf power is needed for the limiting case $t_W \to \tau$. Alternatively, we may say that for a given amount of rf power the minimum value of τ is obtained for $\tau \to t_W$.

Going to this limit means continuous irradiation; however, it also means losing the otherwise convenient observation windows. To overcome this difficulty Mehring and Waugh[107] proposed to construct a large efficient cycle out of a carefully optimized number of "averaging subcycles" with $t_W = \tau$ followed by one "observation cycle" with $t_W < \tau$. They have successfully carried out one experiment of this type and have described some further refinements.[107]

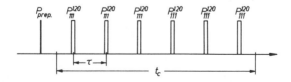

FIG. 5-14. Six-pulse sequence. The rotation is about the rotating frame 111 direction. This sequence is the pulsed analog of the Lee–Goldburg experiment[61] with periodic reversals of the effective field. For experimental details, see Haeberlen and Waugh.[51]

[107] M. Mehring and J. S. Waugh, *Phys. Rev. B* **5**, 3459 (1972).

Interest in this line of search for higher resolution in solids has weakened since it became clear that WAHUHA-type sequences including the MREV sequence also work for finite pulse widths up to $t_W = \tau$. Recall, however, that going to the limit is not advisable for reasons of compensation of the rf inhomogeneity (see Section D,1,c).

2. SYMMETRIZING

We have seen that multiple-pulse sequences can be designed in such a way that not only the average dipolar Hamiltonian $\overline{\mathcal{H}}_D$ vanishes over the cycle, but also the first-order correction term, in fact, all correction terms of odd order. Also we mentioned briefly in Section B that it is possible to design sequences leading to $\overline{\mathcal{H}}_D = \overline{\mathcal{H}}_D^{(1)} = \overline{\mathcal{H}}_D^{(2)} = 0$.

The first sequence of this type has been proposed by Evans.[94] His cycle is shown in Fig. 5-15. One of its characteristic features is to have three different time intervals.

Mansfield[83] proposed an alternative solution based on "symmetrizing." To understand what "symmetrizing" means, consider the WAHUHA sequence again. In the toggling frame the dipolar Hamiltonian runs through the following sequence of states:

$$\text{(a)}\quad \mathcal{H}_D^{\ z}\ \ \mathcal{H}_D^{\ y}\ \ \mathcal{H}_D^{\ x}\ \ \mathcal{H}_D^{\ x}\ \ \mathcal{H}_D^{\ y}\ \ \mathcal{H}_D^{\ z}.$$

The WAHUHA sequence or, as we say now, sequence (a) of Hamiltonian states, produces

$$\overline{\mathcal{H}}_D^{(2)} \propto \left[(\mathcal{H}_D^{\ z} - \mathcal{H}_D^{\ x}), [\mathcal{H}_D^{\ x}, \mathcal{H}_D^{\ y}] \right]$$

[see Eq. (5-1)]. Now suppose we have constructed multiple-pulse sequences that produce the following sequences of Hamiltonian states:

$$\text{(b)}\quad \mathcal{H}_D^{\ x}\ \ \mathcal{H}_D^{\ z}\ \ \mathcal{H}_D^{\ y}\ \ \mathcal{H}_D^{\ y}\ \ \mathcal{H}_D^{\ z}\ \ \mathcal{H}_D^{\ x},$$

$$\text{(c)}\quad \mathcal{H}_D^{\ y}\ \ \mathcal{H}_D^{\ x}\ \ \mathcal{H}_D^{\ z}\ \ \mathcal{H}_D^{\ z}\ \ \mathcal{H}_D^{\ x}\ \ \mathcal{H}_D^{\ y}.$$

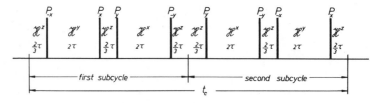

FIG. 5-15. Evans' eight-pulse cycle leading to $\overline{\mathcal{H}}_D = \overline{\mathcal{H}}_D^{(1)} = \overline{\mathcal{H}}_D^{(2)} = 0$. We have drawn Evans' eight-pulse cycle in this figure in a symmetric form; therefore, we have in addition $\overline{\mathcal{H}}^{(3)} = 0$.

Sequences (b) and (c) derive from sequence (a) by cyclic permutations of the indices x, y, z.

$\mathcal{H}_D^{(2)}$ pertinent to sequences (b) and (c) clearly derives from Eq. (5-1) by cyclic permutations of the indices x, y, z. The sum of dipolar second-order correction terms pertinent to sequences (a), (b), and (c) vanishes. Formally, we have encountered exactly the same situation when discussing resonance offset averaging (see Section B).

We now ask the question: What pulse sequences will produce the sequences of Hamiltonian states (b) and (c)? The answer is simple: Variants of the WAHUHA sequence sandwiched between P_y^{90}, P_{-y}^{90} and P_{-x}^{90}, P_x^{90} pulses (see Fig. 5-16).

A combination of sequences (a), (b), and (c) gives a 16-pulse supercycle, which obviously has the same zero-order properties as the WAHUHA sequence. $\mathcal{H}_D^{(1)} = 0$ as before, but in addition we now have $\mathcal{H}_D^{(2)} = 0$. Because $\mathcal{H}^{(2)}$ is symmetric in $\tilde{\mathcal{H}}(t_3)$ and $\tilde{\mathcal{H}}(t_1)$, one can actually achieve the same result with a somewhat simpler 14-pulse sequence.[83]

In order to kill the next correction term $\mathcal{H}_D^{(3)}$, and, in fact, again all correction terms of odd order, one may combine variants of such supercycles consisting of different permutations of (sandwiched) WAHUHA-type cycles in such a way as to end up with symmetric hypercycles having the property

$$\mathcal{H}_D = \mathcal{H}_D^{(1)} = \mathcal{H}_D^{(2)} = \mathcal{H}_D^{(3)} = 0.$$

The same game can be played using as basic buildings blocks the MREV eight-pulse cycle, or even a series of MREV cycles some of which include 270° pulses.

These "games" are, at least at present, still academic for a number of

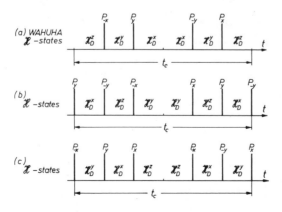

FIG. 5-16. Pulse sequences producing cyclicly permuted Hamiltonian states.

reasons. Two experimental reasons are

(i) The high complexity of the sequences; we point out, however, that generating these sequences electronically is not a serious obstacle.

(ii) The scarcity of observation windows. This drawback is not severe with single line spectra, but it is very severe with complex spectra extending over an appreciable range of frequencies.

One of the open theoretical questions is finite pulse widths. Recall that the design of the super- and hypercycles was based on infinitely short rf pulses.

VI. Applications of Multiple-Pulse Techniques

A. Which Nuclei Are Candidates?

The stated goal of high-resolution NMR in solids by multiple-pulse techniques is the measurement of shielding tensors in the presence of strong homonuclear dipolar couplings. For what nuclei—to be modest let us add with $I = \frac{1}{2}$—do we typically encounter in solids the situation invoked by the preceding sentence? We start by considering

1. Protons

In typical organic compounds and inorganic compounds containing crystal water the proton dipolar linewidth is of the order of 20 to 60 kHz. The splitting of the proton resonance from, e.g., isolated CH_2 groups and H_2O molecules ranges up to 64 and 82.5 kHz, respectively. Proton chemical shifts, including shift anisotropies, cover a range of about 30 ppm. Recall that isotropic chemical shifts of protons in diamagnetic compounds cover a range of roughly 12 ppm only, aside from a few pathological exceptions. 30 ppm corresponds to line shifts of 1.8, 2.7, and 10.8 kHz, respectively, for 60, 90, and 360 MHz spectrometers, which are or are becoming standard.

The conclusion is that proton shifts in solids are usually masked by dipolar line broadening even for 90- and 360-MHz spectrometers. Multiple-pulse techniques are definitely required to measure proton shielding anisotropies. This is particularly so as the results of the liquid-crystal technique—practically the only alternative source of information for proton shielding anisotropies—are such that the authors of a very recent review[30] feel that they must qualify them as "generally unreliable."

The numbers quoted above make clear, on the other hand, that meaningful multiple-pulse work on protons in solids can be done with a spectral resolution much worse than is standard in high-resolution NMR in liquids. Linewidths in excess of 200 Hz are often tolerable, provided the spectra are simple enough in the sense that they contain a few lines only.

It should also be mentioned that there are compounds where the protons or pairs of protons are so diluted that a straightforward measurement of proton shielding tensors is possible. An example is trichloroacetic acid.[108]

The next candidate for multiple-pulse work is

2. ^{19}F

Dipolar linewidths in completely fluorinated compounds tend to be somewhat smaller than those of the corresponding protonated compounds; the main reason is larger ^{19}F–^{19}F internuclear distances. The largest dipolar splitting of the pair of lines from, e.g., a CF_2 group is 33 kHz.

A major difference from protons exists in the range of chemical shifts and shift anisotropies. Typical values of ^{19}F shift anisotropies are 150 ppm, but numbers in excess of 1000 ppm have been reported (see Tables X–XIII of Appleman and Dailey[30]). 150 ppm corresponds to lineshifts of 9, 13.5, and 54 kHz for 60-, 90-, and 360-MHz spectrometers, respectively. Hence, the range of ^{19}F chemical shifts (shift anisotropies) is comparable to or even larger than ^{19}F–^{19}F dipolar line broadening—provided one exploits the strongest magnetic fields now available for NMR purposes. Modest suppression of ^{19}F–^{19}F dipolar couplings by some multiple-pulse sequence will already reveal ^{19}F shift anisotropies in simple cases. Quite naturally, more complex compounds with more complex spectra become amenable to investigation if higher resolution is available.

Skipping ^{3}He (exotic) and ^{203}Tl, ^{205}Tl (exotic, strongly toxic), which would come next according to their magnetic moments, we proceed to consider briefly

3. ^{31}P

On the one hand ^{31}P chemical shift anisotropies are usually large—several hundreds of ppm (see Table[109] XIV of Appleman and Dailey[31]). On the other hand the density of ^{31}P in solids is usually so low that in reasonably strong magnetic fields the spread of the spectra due to chemical shifts by far exceeds the ^{31}P–^{31}P dipolar line broadening. An example is provided by Fig. 6-1, which shows the ^{31}P spectrum of solid white phosphorous recorded at 25 K by standard Fourier transform techniques at 92 MHz. This frequency

[108] D. C. Haddix and P. C. Lauterbur, *Nat. Bur. Stand. (U.S.), Spec. Publ.* **301**, 403 (1969).
[109] The value of $\Delta\sigma$ quoted there for P_4 is in error; see below and Spiess *et al.*[37]

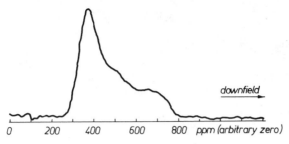

FIG. 6-1. ^{31}P spectrum of solid white phosphorous; $v_0 = 92$ MHz, $T = 25$ K, $\Delta\sigma = -405 \pm 15$ ppm (from Speiss *et al.*[37]).

corresponds to a magnetic field of "only" 5.3 Tesla (53 kG)—8.5 Tesla is becoming standard!

White phosphorous consists of P_4 tetrahedra; the ^{31}P nuclei sit on threefold (molecular) symmetry axes. The ^{31}P shielding tensors are, as a result, axially symmetric. This is what Fig. 6-1 displays. The anisotropy is $\sigma_\parallel - \sigma_\perp = -405$ ppm. This corresponds at 92 MHz to more than 37 kHz. According to a computer fit of this spectrum, but also according to a theoretical estimate based upon the second moment, the dipolar width of the component lines is about 2.5 kHz.

^{31}P spectra of single crystals of P_4S_3 have been recorded at 98 MHz by Gibby *et al.*[110] Here, too, the chemical shifts by far exceed the dipolar linewidths.

These examples are typical, and only in complex cases where very high resolution is really required will there be a need of bothering about multiple-pulse techniques if one is interested in ^{31}P shielding anisotropies. Usually it is much simpler and eventually also much cheaper to buy a high-field superconducting solenoid and to do just standard FT spectroscopy.

What we have said about ^{31}P holds true, a fortiori, of all further nuclei with spin $\frac{1}{2}$. Thus our final conclusion is:

For the measurement of chemical shift tensors, multiple-pulse line-narrowing techniques are really needed only for protons, and to a lesser extent already, for ^{19}F.

This is not to say, however, that there are no applications at all of multiple-pulse techniques to other nuclei.

Mehring and Raber[111] have applied, for example, such techniques to powder samples of metallic ^{27}Al $(I = \frac{5}{2})$ and ^{9}Be $(I = \frac{3}{2})$ and obtained highly precise values for the isotropic Knight shifts.

[110] M. G. Gibby, A. Pines, W.-K. Rhim, and J. S. Waugh, *J. Chem. Phys.* **56**, 991 (1972).
[111] M. Mehring and H. Raber, *Proc. 1st Spec. Colloq. AMPERE, 1973* p. 216 (1973).

B. Applications to [19]F

Applications of multiple-pulse line-narrowing techniques to ^{19}F have been reviewed very recently by Appleman and Dailey in an article about nuclear magnetic shielding and bulk susceptibility anisotropies.[30] This article is recommended for information concerning

(i)   alternative experimental approaches to chemical shift anisotropies such as liquid-crystal solvent, molecular-beam, and double-resonance techniques, and relaxation and second moment methods, etc;

(ii)   a comprehensive list of, on the one hand ab initio calculated, and on the other hand, experimentally measured ^{19}F shift tensors and shift anisotropies including data published until 1972;

(iii)   the current status of ab initio calculations of nuclear magnetic shielding tensors.

We shall not undertake to review this field again; rather we shall restrict ourselves to some representative applications of multiple-pulse techniques.

1. A "Historical" Example: The First Measurement of Chemical Shifts in a Solid by Multiple-Pulse Techniques[112]

In the MIT NMR group we have been able since the end of 1967 to narrow down to 100–200 Hz the ^{19}F resonance line in CaF_2 by using the WAHUHA sequence, but it was not before the fall of the following year that we could observe for the first time a beat structure in the multiple-pulse response of a solid sample (see Fig. 6-2) and, correspondingly, a spectrum with resolved

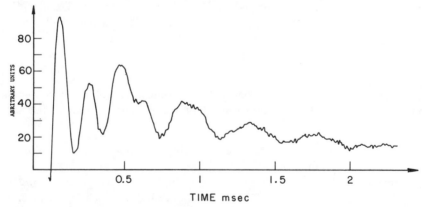

FIG. 6-2. NMR signal of TFE/PFMVE during a WAHUHA experiment. The train contained about 100 four-pulse cycles (from Ellett et al.[112]).

[112] D. Ellett, U. Haeberlen, and J. S. Waugh, *Polym. Lett.* 7, 71 (1969).

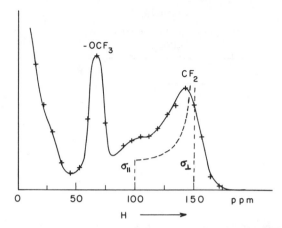

FIG. 6-3. Fourier transform of the trace shown in Fig. 6-2 (from Ellett *et al.*[112]).

peaks (Fig. 6-3). The sample happened to be a tetrafluoroethylene/perfluoro-methylvinyl ether (TFE/PFMVE) copolymer. We were able to determine the (isotropic) ^{19}F chemical shifts in the $-OCF_3$ and CF_2 positions and made an attempt to derive a value for the ^{19}F shift anisotropy in the CF_2 position. The result ($\approx +50$ ppm), however, is not very meaningful because of ill-defined molecular motions. A breakthrough with respect to shift anisotropies was achieved in 1970 by a series of

2. POWER STUDIES OF PERFLUORINATED COMPOUNDS

These studies were carried out by the MIT NMR group.[14] Textbook examples of powder patterns characteristic of both axially symmetric (C_6F_6; Fig. 2 of Mehring *et al.*[14]) and nonaxially symmetric (Fluoranil, $C_6F_4O_2$; Fig. 1 of Mehring *et al.*[14]) shielding tensors were presented. The authors were able to determine ^{19}F principal shielding components in a number of compounds with unprecedented unambiguity. It is worthwhile to stress this point since the bulk of attempts to measure shielding anisotropies (a less ambitious goal than measuring principal shielding components!) mainly by the liquid-crystal solvent technique is beset with a frustrating amount of ambiguities. In fact, the discussion of ^{19}F shielding anisotropies by Appleman and Dailey[30] is primarily concerned with such questions.

The most interesting features of the results of ^{19}F multiple-pulse powder studies are

(a) ^{19}F shielding tensors are not generally axially symmetric as has often been assumed previously. The principal components of the traceless constituent $\sigma^{(2)}$ of σ in, e.g., Teflon $[CF_2]_n$ are, respectively, -80, $+21$, and $+59$ ppm.

TABLE 6-1

COMPARISON OF ^{19}F PRINCIPAL SHIELDING COMPONENTS IN CF$_3$ GROUPS AND AROMATIC SYSTEMS[a]

	$\sigma_{xx}^{(2)}$	$\sigma_{yy}^{(2)}$	$\sigma_{zz}^{(2)}$	Ref.
(CF$_3$CO)$_2$O ($T = 40$ K)	-75	-1	$+76$	[14]
CF$_3$COOAg[b] ($T = 83$ K)	-67	-3	$+70$	[19]
Hexafluorobenzene	-51.7	-51.7	$+103.4$	[14]
Perfluoronaphthalene	-50	-50	$+100$	[14]
Perfluorobenzophenone	-66	-31	$+97$	[14]
Perfluorobiphenyl	-45	-45	$+90$	[14]
Potassium tetrafluorophthalate	-64	-48	$+89$	[113]

[a] The table gives the principal components of $\boldsymbol{\sigma}^{(2)}$ which is the traceless symmetric constituent of $\boldsymbol{\sigma}$.

[b] A single-crystal crystal study of this compound has been carried out by Griffin et al.,[19] see also Section B,3.

TABLE 6-2

PRINCIPAL ^{19}F SHIELDING COMPONENTS FOR FLUORINATED BENZENES[a,b]

[a] After Mehring et al.,[77] p. 35.

[b] The assignment of the principal direction is according to the results of a single-crystal study of perfluorophthalate.[113]

[113] R. G. Griffin, H.-N. Yeung, M. D. LaPrade, and J. S. Waugh, J. Chem. Phys. 59, 777 (1973).

(b) ^{19}F principal shielding components in related compounds tend to be similar. As we shall see presently this is not a matter of course. Examples for CF$_3$ groups and aromatic systems are given in Table 6-1.[19, 14, 113]

(c) There are striking differences in the ^{19}F principal shielding components in compounds that are supposedly closely related (see Table 6-2).

The conclusions one is pushed to draw from Table 6-2 are as follows: Each *ortho*-fluorine shifts the *zz* component by about $+50$ ppm. Para and meta substitution does not affect σ_{zz} appreciably. σ_{yy} (bond direction) is rather insensitive to any kind of substitution and σ_{xx} changes gradually. Clearly, there is also a change in the mean or isotropic value of the shift, which has been known for some time as the so-called ortho effect.[114, 115] So far there is no theory that satisfactorily accounts for the ortho effect.

(d) The quality of the experimental results is so good that asking the following question is not only meaningful but urgent: What do the measured principal shielding components tell us about the compounds, and possibly about their electronic structures? After all, one of the motivations for measuring shielding tensors is getting information about electronic states!

Mehring *et al.* have tried to analyze some of their data in the framework of valence theory,[114] which relates the so-called paramagnetic parts (see Section C,3) of the principal shielding components to the "populations" of the fluorine p orbitals. These populations, in turn, are related to the amounts of double bond (ρ) and ionic (I) characters, and to the degree of sp hybridization (s) of the C—F bond. The numbers quoted by Mehring *et al.*[14] for $(I + s - I \cdot s)$ imply that the C—F bonds in hexafluorobenzene, perfluoronaphthalene, fluoranil, teflon, etc., are either almost purely ionic or almost purely double bonds. Both implications are evidently unacceptable. The main conclusion is thus that valence theory in the formulation of Karplus and Das[114] is inadequate for describing ^{19}F principal shielding components. Let us try to understand why this is so. First, there is the difficulty of separating the experimental quantities σ_{xx}, σ_{yy}, σ_{zz} into their "paramagnetic" and "diamagnetic" parts. The way Mehring *et al.* solve this difficulty is reminiscent of the origin of the zero of the Fahrenheit temperature scale.[116] They assume that the most shielded value observed until 1971 (σ_{zz} of C$_6$F$_6$) is due solely to diamagnetic shielding ($\sigma^{(d)}$) and that this value can be taken as a fair approximation of $\sigma^{(d)}$ also for the other compounds of the series considered. Second, and much more seriously, the Karplus and Das theory has a built-in inadequacy since it resorts to the so-called closure approximation. The closure approximation

[114] M. Karplus and T. P. Das, *J. Chem. Phys.* **34**, 1683 (1961).

[115] J. Nehring and A. Saupe, *J. Chem. Phys.* **52**, 1307 (1970).

[116] Fahrenheit defined as zero the lowest temperature he could produce by mixing water, ice, and ammonium chloride.

means that one single "average" excitation energy is assigned to all excited electronic states of the molecule.

Now, from the analysis of some ^{13}C shielding tensors[117] we do know that $\sigma_{xx}^{(p)}$, $\sigma_{yy}^{(p)}$, $\sigma_{zz}^{(p)}$, and in particular the differences between them are intimately related to one or at most very few of the excited electronic states and their excitation energies. This means that any theory based on the closure approximation is bound to fail in these cases. The same situation can safely be expected to prevail for ^{19}F and other nuclei in many compounds.

(e) Powder studies in general can reveal only principal shielding components, but not principal directions. When these directions are fixed by symmetry, powder studies still cannot reveal which principal component belongs to which principal direction except when axial symmetry of σ is enforced by crystal symmetry (three- or higher-fold axis through site of nucleus of interest).

A special case is encountered in C_6F_6, where such a symmetry can be switched on and off by changing the temperature of the sample.

The principal ^{19}F shielding directions in C_6F_6 are fixed by molecular symmetry: one is perpendicular to the molecular plane and thus parallel to the molecular C_6 axis, another is parallel to the C—F bond, which is a C_2 axis, and the third is necessarily perpendicular to the other two. At 200 K Mehring et al.[14] obtained from C_6F_6 a powder pattern characteristic of an axially symmetric shielding tensor. This is no surprise since the C_6F_6 molecules are known to reorient rapidly about their sixfold axes at this temperature. The parallel component $\sigma_\parallel$ (shoulder of powder pattern) is a principal shielding component and necessarily corresponds to the C_6 axis. $\sigma_\perp$ (peak of the powder pattern) is, on the other hand, the average of the two in-plane components. The experiment was repeated at 40 K, where the reorientations have slowed down enough so that the molecules may be considered—with regard to shielding—as being stationary. Surprisingly, the 40 K multiple-pulse spectrum turned out to be identical to the 200 K spectrum. This implies that the in-plane shielding components are equal (to within the accuracy of the experiment) and that the ^{19}F shielding tensor in C_6F_6 is axially symmetric. However, the symmetry axis is not the bond direction, but rather the sixfold symmetry axis of the molecule. Note that the ^{19}F site symmetry in C_6F_6 does not include a sixfold symmetry axis so that this result is not a consequence of symmetry. C_6F_6 is an exceptional case, however.

In general, principal shielding directions can only be found by

3. SINGLE-CRYSTAL STUDIES

^{19}F shielding tensors have been studied by multiple-pulse techniques in single crystals of the following compounds:

[117] J. Kempf, H. W. Spiess, U. Haeberlen, and H. Zimmermann, Chem. Phys. 4, 269 (1974).

CaF_2: not interesting because the ^{19}F site symmetry is cubic; MgF_2, ZnF_2: Stacey et al.[118], Vaughan et al.[119]; silver trifluoro-acetate: Griffin et al.[19]; potassium tetrafluorophthalate: Griffin et al.[113]

a. MgF_2, ZnF_2[118,119]

MgF_2 and ZnF_2 are isomorphic. The site symmetry of ^{19}F is C_{2v}; as a consequence all three principal directions are determined by symmetry. What remains for the experimenter is to measure σ_{xx}, σ_{yy}, σ_{zz} and to assign them to "their" principal directions.

The first part of the task can be done in a straightforward way by taking rotation patterns about suitable axes (see Chapter III, Section B). This is what Stacey et al.[118] and Vaughan et al.[119] did. Their results for σ_{xx}, σ_{yy}, and σ_{zz} are 13, 28, and 43 ppm for MgF_2, and 15, 38, and 59 ppm for ZnF_2 relative to liquid hexafluorobenzene.

The second part involves a not unusual difficulty: While the data allow an unambiguous assignment of σ_{xx} to the C_2 axis, σ_{yy} and σ_{zz} cannot be assigned unambiguously. The reason is that in MgF_2 (and in ZnF_2) there are two magnetically inequivalent but crystallographically equivalent fluorine sites, and it so happens that the y direction of one site coincides with the z direction of the other. As there is no purely experimental way of telling which of the (in general) two lines of the spectra arises from which site there is no purely experimental way of assigning σ_{yy} and σ_{zz}. Vaughan et al.[119] do propose an assignment, but it is based on theoretical (which means on completely different) grounds than the assignment of σ_{xx}.

b. Silver Trifluoro-Acetate[19]

The trifluoromethyl group in silver trifluoro-acetate (CF_3COOAg) has been studied by the MIT NMR group both at room temperature and at 40 K. At room temperature the CF_3 group is rapidly reorienting about the C—C bond, so partially averaged shielding tensors can be measured only. At 40 K the reorientations are effectively frozen out. What makes this study of the CF_3 group particularly exciting is the following:

In CF_3COOAg kept at 40 K the F atoms of the CF_3 group are magnetically inequivalent. In fact, there are two magnetically inequivalent but crystallographically equivalent CF_3 groups. This is, however, not so important and therefore let us focus attention on one CF_3 group. Its F atoms are not only magnetically but also crystallographically inequivalent. Moreover, there is no symmetry argument whatsoever that would allow a prediction of any one of the principal shielding directions of any one of the F atoms. In short, the shielding tensors of the F atoms in the CF_3 group of CF_3COOAg are completely independent and nothing is predetermined by symmetry.

[118] L. M. Stacey, R. W. Vaughan, and D. D. Elleman, Phys. Rev. Lett. 26, 1153 (1971).

[119] R. W. Vaughan, D. D. Elleman, W.-K. Rhim, and L. M. Stacey, J. Chem. Phys. 57, 5383 (1972).

Of course, our chemical and physical intuition tells us that these shielding tensors ought to be very similar. After all, in an isolated —CF_3 group the F atoms are crystallographically equivalent and we suspect that incorporating CF_3 groups in $(CF_3COOAg)_2$ dimers and into a rather loosely packed crystal does not alter the shielding tensors dramatically.

Now let us look at the experimental results of Griffin et al.[19] At 40 K the following principal shielding components (relative to CaF_2) and anisotropies were measured for the fluorine atoms labeled F1, F2, F3 in Griffin et al.[19]:

	σ_{xx} (ppm)	σ_{yy} (ppm)	σ_{zz} (ppm)	$\triangle \sigma = \sigma_{zz} - \frac{1}{2}(\sigma_{xx}+\sigma_{yy})$ (ppm)
F1	− 64.3	−4.3	+ 60	94.3
F2	− 75.0	+ 1.9	+ 73.1	109.7
F3	− 73.6	− 6.6	+ 80	120.1

The most shielded directions lie approximately along the respective CF bonds, the tilt being 9.7, 11.5, and 7.8° for atoms F1, F2, and F3, respectively. The least shielded directions are approximately perpendicular to the respective C—C—F planes.

We see that there are differences of up to 20 ppm in the principal components and differences in the anisotropy of more than 25 ppm out of a total range of 120 ppm and we really think that such differences should be called dramatic.

Why are these differences so big? Honestly, we do not know, but Griffin et al. at least indicate what they are probably due to. The most conspicuous differences are between F1 on the one, and F2, F3 on the other hand. Griffin et al. suggest that the shielding differences are the result of an eclipsed position of F1 relative to one of the oxygen atoms of the carboxyl group. Explanations on a "higher" level still belong to the land of dreams at present!

Another incentive for investigating CF_3 groups in rigid solids is the following: Deeply engraved in the mind of almost every chemist is the notion that J coupling within CF_3 groups is unobservable. However, the condition for the unobservability of J coupling between F atoms—magnetic equivalence of F atoms—is not fulfilled in a rigid solid and therefore J coupling becomes, in principle, observable. Nevertheless, in this particular experiment it could not be observed, probably because the size of the coupling constants was too small for the available spectral resolution.

c. Potassium Tetrafluorophthalate[113]

In isolated molecules of monofluoro-, 1,3,5-trifluoro-, and hexafluorobenzene all principal shielding directions are fixed by molecular symmetry. Only one is fixed in, e.g., 1,2-difluorobenzene. It is perpendicular to the molecular plane. What about the other two? The results on the trifluoromethyl group quoted above prompt us to be cautious in merely making assumptions.

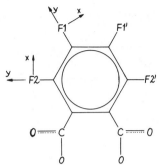

FIG. 6-4. The tetrafluorophthalate anion. Principal shielding components in ppm relative to liquid C_6F_6 are

$$\sigma_{xx}(1) = -58, \qquad \sigma_{yy}(1) = -47, \qquad \sigma_{zz}(1) = +94.5,$$

$$\sigma_{xx}(2) = +73, \qquad \sigma_{yy}(2) = -55, \qquad \sigma_{zz}(2) = +63.$$

After Griffin et al.[113]

Single crystals of fluorinated benzenes are difficult to investigate because of low melting points and complex crystal structures. Potassium tetrafluorophthalate, on the other hand, is convenient and has a bearing on the question just asked: melting point above room temperature and all molecules magnetically equivalent. Griffin et al. studied this compound and found for both types of fluorine (F1 and F2 in Fig. 6-4) that indeed the perpendicular to the molecular plane and the respective C—F bonds are very close to principal ^{19}F shielding directions. This fixes the in-plane perpendiculars to the bonds as third principal directions. For F1 as well as for F2 the perpendicular to the plane is the most shielded, the C—F bond direction the intermediate, and the in-plane perpendicular to the bond the least shielded direction. These results justify the assignments in Table 6-2 for the di- and tetrafluorinated benzenes.

We close our discussion of ^{19}F multiple-pulse single-crystal studies with a comment.

d. Comment on Powder Studies

With few exceptions powder studies can yield accurate values of principal shielding components only when all nuclei that contribute to the spectra occupy crystallographically equivalent sites. We can analyze powder patterns only in these cases in terms of one single set of principal components.

Now, there are very few "systems" in nature where ^{19}F nuclei meet this condition. Thus very few substances appear to be amenable to powder studies. Facing this situation one may be inclined to loosen somewhat the rigor and assume that all nuclei occupying equivalent positions in the isolated molecule or even all nuclei bonded similarly (as are the F atoms in potassium tetrafluorophthalate) have similar enough principal shielding components for the powder pattern to be analyzable in terms of a single set of principal shielding components.

This assumption had been made tacitly in some of the powder studies quoted in Section B,2. One lesson the available single-crystal studies teaches us is—to express it cautiously—that we must be very cautious with such assumptions and that we must be prepared to be misled.

4. MOLECULAR MOTIONS STUDIED BY MULTIPLE-PULSE TECHNIQUES

The traditional way of studying molecular motions in solids by NMR is by spin–lattice relaxation of the Zeeman energy (T_1), of the Zeeman energy in the rotating frame ($T_{1\rho}$), and of the dipolar energy (T_{1D}), and by the second moment of resonance lines. These methods are not very specific to various kinds of motions. At each temperature and spectrometer frequency one gets just a one-number answer from the system.

Powder pattern lineshapes, on the other hand, can be considerably more informative: At each temperature one measures an entire function $g(\omega)$. One condition for $g(\omega)$ to carry specific information about molecular motions is that it be governed by predominantly intramolecular or even single-bond effects such as chemical shift anisotropy in contradistinction to dipolar effects, which are typically almost equally inter- and intramolecular in origin and which rarely are specific to the orientations and motions of a molecule. What we have just stated is borne out by some recent studies of molecular motions where either "high" magnetic fields of about 6 $T \triangleq$ 60 kG are exploited to produce, in absolute terms, large chemical shifts and shift anisotropies of ^{31}P and ^{13}C,[37] or where use has been made of the feasibility to decouple ^{13}C from protons in solids.[120]

An example where ^{19}F line narrowing by multiple-pulse techniques was successfully used to produce the required spectral resolution for studying molecular motions is perfluorocyclohexane,[121,122] C_6F_{12}.

Figure 6-5 shows the ordinary (a) and the multiple-pulse (b) spectrum of C_6F_{12} at 200 K. The multiple-pulse spectrum is approximately an AB quartet reflecting the presence of distinguishable interacting axial and equatorial fluorines ($\delta = 18.2$ ppm, $J = 284$ Hz). At 200 K the multiple-pulse spectrum carries no information about chemical-shift anisotropy. This is due to rapid reorientations of the C_6F_{12} molecules (which are globular in shape) at their lattice sites. Dipolar interactions are reduced by these reorientations, but they are not suppressed altogether. In particular, some of the intermolecular dipole–dipole interactions remain static and lead to the broad structureless spectrum (a) of Fig. 6-5.

Axial and equatorial fluorines are exchanged by chair-to-chair inversions of the C_6F_{12} molecules. High inversion rates as they occur at $T \geqslant 267$ K

[120] A Pines, M. G. Gibby, and J. S. Waugh, *J. Chem. Phys.* **59**, 569 (1973).
[121] D. Ellett, U. Haeberlen, and J. S. Waugh, *J. Amer. Chem. Soc.* **92**, 411 (1970).
[122] D. Ellett, R. G. Griffin, and J. S. Waugh, *J. Amer. Chem. Soc.* **96**, 345 (1974).

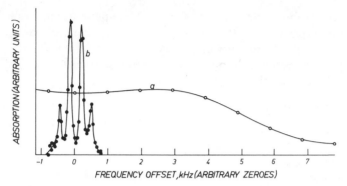

FIG. 6-5. ^{19}F NMR spectra of solid C_6F_{12} at 200 K. (a) Normal spectrum, from Fourier transformation of FID, showing restricted molecular rotation *in situ*; (b) multiple-pulse spectrum under the same conditions, showing lack of ring inversion. From Ellett *et al.*[121] Reprinted with permission from *J. Amer. Chem. Soc.* **92**, 411 (1970). Copyright by the American Chemical Society.

lead to a collapse of the quartet into a single line.

A study of the transition with temperature of the four-line into a single-line spectrum by Ellett *et al.*[122] allowed a measurement of the temperature dependence of the inversion rate and a determination of the associated activation enthalpy and entropy.

C. Applications to Protons

For about two decades, isotropic chemical shifts of protons have become a standard tool of organic chemists, and thousands and thousands of proton "shifts" have been measured with fantastic precision. On the other hand, our knowledge of the anisotropy of proton shifts, or more generally, of proton shift tensors was very poor until very recently.

In their 1974 review of magnetic shielding anisotropies Appleman and Dailey[30] described the situation in the following terms:

"(i)   The most extensively used technique for determining proton anisotropies, the liquid-crystal method, has been found to be generally unreliable due to large uncertainties and the small range of proton shifts.

(ii)   Other techniques have also been less well suited for accurate measurements of proton anisotropies.

(iii)   The ab initio theoretical methods have generally been considerably less successful for proton shielding than for the other nuclei listed above (i.e., for ^{13}C, ^{19}F, $^{14,15}N$, ^{17}O, ^{31}P)."

By "other techniques" Appleman and Dailey evidently mean, in particular, multiple-pulse techniques. As far as we can see, there has been no breakthrough in the liquid-crystal method since the completion of Appleman

and Dailey's manuscript. The success of multiple-pulse techniques in eluci-
dating proton chemical shift tensors, however, has changed markedly and
hopefully the following pages will convince the reader of this claim. Also, at
least a qualitative rationalization of the experimental results is often easier
for protons than for other nuclei such as ^{13}C and ^{19}F.

1. A FIRST SURVEY OF PROTON SHIFT ANISOTROPIES: POWDER STUDIES

Isotropic chemical shifts of protons are characteristic of the chemical bonds
in the immediate neighborhood of the respective protons. This is proven by
the very fact that charts of proton shifts can be constructed. It may be expected
that there is also a relation between bonds and the tensor properties of shifts.
We therefore looked for prototype compounds containing protons in "typical"
aliphatic-, aromatic-, olefinic-, carboxylic-, ..., bonding situations, which, in
addition, were well suited for powder multiple-pulse work.

The most important conditions for being "well suited" are

(i) All protons should possibly be crystallographically equivalent. At
least, they should be chemically equivalent, which means equivalent in the
isolated molecule (see above).

(ii) The compound should preferably be a (rigid) solid at room temperature. This is not merely a matter of convenience as it may seem. For proton
multiple-pulse work it is vital to have the spectrometer alignment at its
optimum. The alignment, however, depends on the temperature of the probe.
Each change of temperature requires realignment. "Interesting" samples
usually cannot be used for realignment because of long relaxation times. Test
samples are therefore needed for realignment and the rapid exchange of test
and "interesting" samples at, say, liquid nitrogen temperatures does cause
complications.

From the point of view of comparing eventual experimental results with
theoretical calculations it would be desirable to choose those very small,
highly symmetric molecules preferred by "ab initio-ians." Unfortunately
most of these compounds—CH_4, C_2H_4, C_2H_6, COH_2, etc.—are either liquids
or gases at room temperature and thus do not qualify as "well suited."

We therefore choose succinic acid anhydride, maleic acid anhydride,
ferrocene, and oxalic acid as prototype compounds containing aliphatic,
olefinic, aromatic, and carboxylic protons, respectively. The protons in all
these compounds are equivalent at least in the isolated molecules and thus meet
condition (i). With the exception of ferrocene they are effectively rigid at room
temperature. The C_5H_5 rings of ferrocene reorient rapidly around their C_5
axes[123] so only the shielding component parallel to this axis ($\sigma_\parallel$), and the
average of the in-plane components are measurable.

Figure 6-6 (reproduced from Haeberlen and Kohlschütter[123]) shows

[123] U. Haeberlen and U. Kohlschütter, *Chem. Phys.* **2**, 76 (1973).

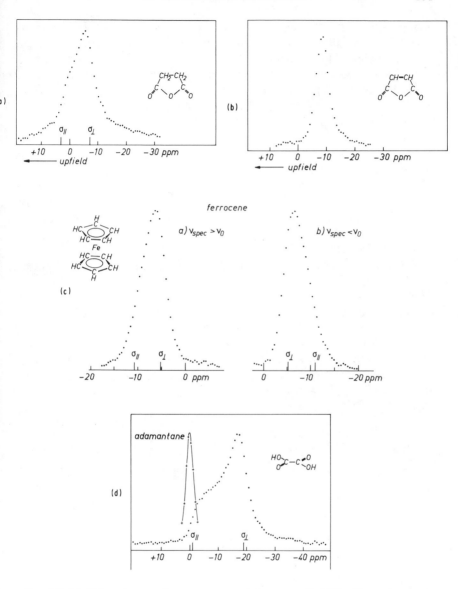

FIG. 6-6. Multiple-pulse powder spectra recorded at $v_0 = 90$ MHz. The reference is always adamantane. (a) Succinic acid anhydride, $\Delta\sigma \approx +10$ ppm; (b) maleic acid anhydride, no anisotropy detectable; (c) ferrocene, both up- and downfield spectra are shown; $\Delta\sigma \approx -5.5$ ppm; (d) anhydrous oxalic acid, $\Delta\sigma = +17.8$ ppm. From Haeberlen and Kohlschütter.[123]

multiple-plus powder patterns of these compounds. None of them is reminiscent of a nonaxially symmetric shift tensor, although our results from single-crystal studies of maleic acid and monopotassium maleate strongly suggest that under the powder spectrum of maleic acid anhydride there is hidden a shielding tensor with $\eta \approx 1$. However, the range of shifts is too small to produce prominent shoulders in the spectrum.

The spectra of ferrocene are slightly asymmetric, which is indicative of a slight shielding anisotropy. The positions of $\sigma_{\parallel}$ and $\sigma_{\perp}$ in Fig. 6-6 are the result of a least-squares computer fit. (A later single-crystal study of ferrocene yielded $\Delta\sigma = -6.5 \pm 0.1$ ppm; see Spiess et al.[102]). The asymmetry of the succinic acid anhydride spectrum is very marked. According to the computer fit the anisotropy is about $+10$ ppm. The anhydrous oxalic acid spectrum is a beautiful example of a powder spectrum resulting from an axially symmetric shielding tensor. The values for $\sigma_{\parallel}$ and $\sigma_{\perp}$ are given in the figure and have essentially been confirmed by a single-crystal study by Van Hecke et al.[124]

This first survey thus led to a spectrum of quite distinct shift anisotropies, but at this stage one cannot claim, of course, that the results are typical for the respective types of bonds. However, further powder studies on malonic, succinic, maleic, fumaric, and phthalic acids,[123] on phthalic acid anhydride,[123] on butyne-dioic and squaric acids,[125] and on trichloroacetic acid,[126] and a rapidly growing series of single-crystal studies to which we shall turn presently indicate that the coarse features of proton shielding anisotropies outlined above are indeed typical for the respective bonds although a substantial range of anisotropies has been encountered already for most of them.

2. SINGLE-CRYSTAL STUDIES

Table 6-3[78, 79, 102, 104–106, 108, 124, 127–134] gives a summary of proton single-crystal studies that have been conducted to date by multiple-pulse and

[124] P. Van Hecke, J. C. Weaver, B. L. Neff, and J. S. Waugh, J. Chem. Phys. 60, 1668 (1974).
[125] H. Raber, G. Bruenger, and M. Mehring, Chem. Phys. Lett. 23, 400 (1973).
[126] W.-K. Rhim, D. D. Elleman, and R. W. Vaughan, J. Chem. Phys. 59, 3740 (1973).
[127] U. Haeberlen, U. Kohlschütter, J. Kempf, H. W. Spiess, and H. Zimmermann, Chem. Phys. 3, 248 (1974).
[128] R. Grosescu, A. M. Achlama, U. Haeberlen, and H. W. Spiess, Chem. Phys. 5, 119 (1974).
[129] A. M. Achlama, U. Kohlschütter, and U. Haeberlen, Chem. Phys. 7, 287 (1975).
[130] H. Feucht, Diplom Thesis, University of Heidelberg, 1975, unpublished.
[131] T. Terao and T. Hashi, J. Phys. Soc. Jap. 36, 989 (1974).
[132] L. B. Schreiber and R. W. Vaughan, Chem. Phys. Lett. 28, 586 (1974).
[133] H. W. Spiess, H. Zimmermann, and U. Haeberlen, submitted to J. Magn. Reson.
[134] H. M. Vieth, H. W. Spiess, U. Haeberlen, and S. Haussühl, submitted to Chem. Phys. Lett.

other techniques in efforts to determine proton shielding tensors. The results were usually obtained by analyzing rotation patterns.

We proceed by commenting on the naked-number results compiled in Table 6-3. Theoretical and physical interpretations are attempted in the following subsections.

a. *Methylene Protons; Magic-Angle Orientation Technique*

The only methylene group studied so far in single crystals is that of malonic acid, $CH_2(COOH)_2$.[104,127] The accurate measurement of CH_2 proton shielding tensors is hampered by two circumstances:

(i) The anisotropy is small—this we know from powder studies.[123]

(ii) As a consequence of the small proton–proton distances in CH_2 groups (≈ 1.76 Å) the dipolar coupling is particularly large.

Let us designate by $\Delta\omega_{HH}$ the proton NMR splitting of the CH_2 group. The size of the product $\tau\Delta\omega_{HH}$ gives an idea of the difficulty of suppressing the respective interactions by multiple-pulse sequences. The smallest value of τ at which we can currently operate our spectrometer is 2.4 μsec; $\Delta\omega_{HH} = 2\pi \times 64$ kHz for B_{st} parallel to the H—H internuclear vector r_{HH}. The product $\tau\Delta\omega_{HH}$ is therefore close to unity, which means that a good spectral resolution cannot be obtained.

Malonic acid, however, also offers a favorable circumstance: Its crystal structure is such that all molecules are magnetically equivalent. In a single crystal the vectors r_{HH} of all molecules are parallel. Therefore, it is possible to orient the crystal such that the angle between all r_{HH} and B_{st} becomes equal to the magic angle $\vartheta_m = 54°44''$ with the result that the dipolar proton coupling within all CH_2 groups vanishes for purely geometric reasons. As the CH_2 groups in malonic acid are magnetically fairly well isolated from the rest of the crystal, the conditions for a successful multiple-pulse experiment are now very favorable, which is demonstrated by the spectra of Fig. 6-7.

It is important to realize that there is not only one, but a onefold infinity of magic-angle orientations of B_{st} relative to r_{HH}. They lie on a cone around r_{HH} (see Fig. 6-8). Measurements along these favorable orientations yield five relations[104] for the six independent elements of σ ($=\sigma^{(s)}$). Recall that by rotating a crystal about an axis perpendicular to B_{st} information for only three relations can be gathered.

In order to get a sixth relation, Kohlschütter[104] also recorded spectra with B_{st} in the plane perpendicular to r_{HH} (see Fig. 6-8). This plane is a mirror plane for the —C—CH_2—C— fragment of malonic acid. By local symmetry we may expect that the two methylene proton shielding tensors are mirror images with respect to this plane (or at least nearly so) with the result that only one NMR line is observed for B_{st} falling into that plane. This is borne out by

TABLE 6-3

PROTON SHIELDING TENSORS MEASURED IN SINGLE CRYSTALS[a]

Compound	Type of proton	σ_{xx} (ppm)	σ_{yy} (ppm)	σ_{zz} (ppm)	Approximate principal direction		Method and reference
					Most shielded	Least shielded	
$(COOH)_2$	H bond, intermolecular	−19.2	−17.1	−1.5	∥ H bond	—	MP, 54 MHz (124)[b]
$CH_2(COOH)_2$	H bond, intermolecular, I	−23.5	−20.9	−2.5	∥ H bond	⊥ carboxylic plane	MP, 90 MHz (104, 127)[c]
	intermolecular, II methylene	−23.0	−19.0	−2.7	∥ C—H bond	⊥ C—H bond	
HCCOOH	H bond, intermolecular	−23.0	−18.7	−1.8	∥ H bond	⊥ molecular plane	MP, 90 MHz (128)[d]
‖ HCCOOH	intramolecular	−26.7	−21.7	−1.5	∥ H bond	⊥ molecular plane	
	olefinic	−10.6	−7.9	−4.3	∥ C=C double bond	⊥ molecular plane	
HCCOOH	H bond, intramolecular	−32.5	−29.6	−0.7	∥ H bond	⊥ molecular plane	MP, 90 MHz (129)[d]
‖ HCCOOK	olefinic	−9.0	−6.8	−4.1	∥ C=C double bond	⊥ molecular plane	
$MgSO_4 \times H_2O$	H bond	−15.8	−15.8	+3.8	∥ H bond	⊥ H bond	(105)[e]
KHF_2	H bond	−39.7	−32.2	+8.8	∥ H bond	⊥ "anionic" plane	MP plus ^{1}H—^{19}F hetero (79)
$KHCO_3$	H bond, dimers	−26.8	−23.3	+2.0	∥ H bond	⊥ dimeric plane	MP, 90 MHz (130)
$KHSO_4$	H bond, dimers	−21.2	−20.3	+3.5	∥ H bond	⊥ dimeric plane	MP, 90 MHz (106)
	chains	−24.4	−22.8	+3.5			

(CCl₃COOH)₂	H bond, dimers	−19.7	−15.7	+0.3	—	—	straightforward, 100 MHz (108)ᶠ
KH₂PO₄	H bond, intermolecular	−27.1 / −29.7	−27.1 / −29.7	+7.5 / +7.3	∥ H bond	⊥ H bond	(131)ᵍ MP, 60 MHz (78)ʰ
Ca(OH)₂	no H bond!	−9.3	−9.3	+4.7	∥ OH⁻ direction	⊥ OH⁻ direction	MP, 60 MHz (132)ⁱ
Trans-C₂H₂I₂	olefinic	−10.9	−8.4	−6.1	15° from C—H bond	⊥ molecular plane	MP, 90 MHz (133)
Fe(C₅H₅)₂	aromatic	−10.4	−3.9	−3.9	within molecular plane	⊥ molecular plane	MP, 90 MHz (102)ʲ
Ca[HCOO]₂	site 1 / site 2	−14.4 / −14.2	−11.0 / −12.4	−8.3 / −8.1			MP, 90 MHz (134)ᵏ

[a] MP, multiple pulse, either WAHUHA or MREV. All shifts are referenced to TMS. For error limits see original papers.
[b] Axial symmetry assumed for analysis of principal directions.
[c] Data for methylene protons are preliminary.
[d] See text for analysis of data of olefinic protons.
[e] Axial symmetry assumed.
[f] Principal directions not fully determined.
[g] First-moment technique.
[h] Multiple pulse plus ^{31}P heterodecoupling.
[i] Principal directions fixed by crystal symmetry.
[j] Axial symmetry is consequence of molecular reorientations; $\sigma_\perp$ is an average of the "true" in-plane components.
[k] The orientations of the PAS's have been determined with respect to crystallographic axes. Unambiguous assignments to the two types of crystallographically inequivalent protons have not been possible yet and, indeed, are impossible with crystallographic and NMR data alone.

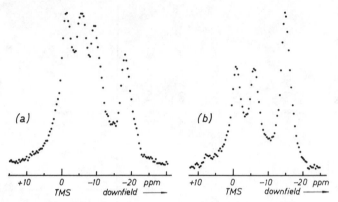

FIG. 6-7. Multiple-pulse spectra of a single crystal of malonic acid, $CH_2(COOH)_2$. (a) General magic-angle orientation, all four possible lines are separated; the two left ones are from the methylene, the two right ones from the carboxylic protons; (b) special magic-angle orientation, $\mathbf{B}_{st}$ within the plane of the CH_2 group (orientation ① in Fig. 6-8). The two carboxylic lines (right) coalesce. The center line probes $\sigma_\perp$ of one, the left line probes almost $\sigma_\parallel$ of the other methylene proton. It is not yet certain, however, whether or not σ(methylene) is axially symmetric about the C—H bond.

Kohlschütter's spectra. The analysis of his data is not complete at the time of this writing but from his spectra it follows directly that the shielding anisotropy of the methylene protons in malonic acid is somewhere between $+5$ and $+6$ ppm. It is thus substantially smaller than in succinic acid anhydride. The

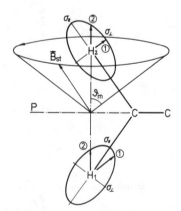

FIG. 6-8. The magic-angle orientation technique applied to a CH_2 group. $\sigma(1)$ and $\sigma(2)$ are probed by moving $\mathbf{B}_{st}$ on the cone around r_{HH} with $\vartheta = \vartheta_m$. For $\mathbf{B}_{st}$ falling into the H—C—H plane (orientation ①), $\sigma_\perp(1)$ is probed, and $\sigma_\parallel(2)$ is probed almost. Provided $\sigma(1)$ and $\sigma(2)$ are mirror images with respect to the plane P, the two methylene NMR lines will coalesce for orientation ② as they will for $\mathbf{B}_{st}$ falling into the plane P.

methylene proton shielding tensors in malonic acid appear to be axially symmetric about the C—H bonds.

Another example where the magic-angle orientation technique has been applied successfully to study the shielding of closely neighbored protons[105] is Kieserit, $MgSO_4 \times H_2O$. Malonic acid and Kieserit are extremely simple cases: Their spectra consist of not more than four and two lines, respectively. More typically, proton spectra consist of many more lines, which often can be neither assigned nor even resolved. Now suppose a sample of interest contains, among other protons, CH_2 groups (or water molecules) that are magnetically well isolated. By orienting a single crystal in such a way that specific CH_2 groups (or water molecules) are at the magic angle to B_{st}, we may hope to "see" these protons as pairs of sharp lines against a broad background. In other cases the sharpness of lines from CH_2 groups that happen to be at the magic angle may help in solving assignment problems.

b. Olefinic Protons

These have been studied so far in single crystals of maleic acid,[128] monopotassium maleate,[129] and trans-diiodoethylene.[133] The outstanding features of their shielding tensors are

(i)   They are not even approximately axially symmetric; in fact, in all three cases the asymmetry parameter η was found to be close to unity.

(ii)   The total anisotropy $\sigma_{zz} - \sigma_{xx}$, is about 6 ppm.

(iii)   The normals to the molecular planes of maleic acid, of its anion, and of trans-diiodoethylene were found to be the least shielded directions.

(iv)   The C—H bond direction is not one of the in-plane principal shielding directions.

For maleic acid and its anion the experimental evidence is that the most shielded direction is along the adjacent C=C double bond. We were essentially led to this conclusion by the fact that we could never see separated lines from the two olefinic protons of individual maleic acid molecules (or maleate anions). Recall that if the C=C double bond direction is a principal shielding direction for the olefinic protons, their σ tensors coincide by molecular symmetry (at least in isolated molecules), and separated lines cannot be observed.

In trans-diiodoethylene we could observe all lines consistent with molecular and crystal symmetry, namely two (see Fig. 6-9). Neither the C—H bond nor the C=C double bond is a principal shielding direction. As we mentioned already, self-decoupling (see Chapter IV, Section F,5,b) is crucial in trans-diiodoethylene for decoupling the protons from the iodines. The self-decoupling efficiency depends strongly on the orientation of B_{st}, and in some orientations it is definitely poor. This is why one of the lines in Fig. 6-9 is so

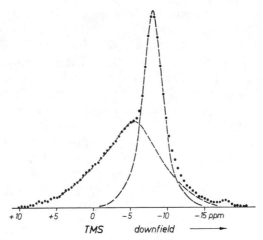

FIG. 6-9. Multiple-pulse spectrum of a spherically shaped single crystal of *trans*-diiodoethylene. B_{st} was oriented such that it was in the plane of all $C_2H_2I_2$ molecules, bisecting the directions of the C=C double bonds of the two kinds of inequivalent molecules. This orientation probes σ_{zz} (broad line) of one, and σ_{yy} (narrow line) of the other kind of molecules. The two lines have been separated visually.

broad. As the orientations for effective and poor self-decoupling are predictable it is possible to tell which of the two lines arises from which of the two magnetically inequivalent molecules of the unit cell, which is crucial for assigning the y and z principal directions. Such a possibility did not exist in the similar cases of MgF_2 and ZnF_2.

c. *Aromatic Protons*[102]

These have been investigated in single crystals of ferrocene, $Fe(C_5H_5)_2$. Spectra were recorded at room temperature only. The most serious reason for not extending the measurements to low temperatures is the impossibility of cooling single crystals of ferrocene down to, say, liquid nitrogen temperature without crushing them. As we have mentioned already, the C_5H_5 rings are rotating rapidly about their five-fold axes at room temperature, so $\Delta\sigma = \sigma_{\parallel} - \sigma_{\perp}$ is all that can be measured from the traceless symmetric part of σ. Our result is $\Delta\sigma = -6.5\pm0.1$ ppm. It is not so much this result and its unusual accuracy that makes ferrocene outstanding, but the chance it offers for studying the interplay between nuclear magnetic shielding and bulk susceptibility in determining the spectral positions of NMR lines. In general orientations of B_{st} there are two ferrocene NMR lines (see Fig. 6-10). Their rotation patterns must lead to axially symmetric shielding tensors with the unique directions along the axes of the two kinds of inequivalent molecules per unit cell. This turns out to be the case only if both the shape and intrinsic anisotropy of the

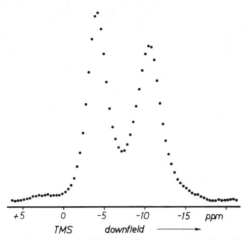

+5   0   -5   -10   -15   ppm
TMS   downfield ⟶

FIG. 6-10. Multiple-pulse spectrum of a spherically shaped single crystal of ferrocene. B_{st} was approximately parallel to the axes of one, and approximately within the planes of the C_5H_5 rings of the other kind of inequivalent molecules. Effectively, $\sigma_\parallel$ (right) and $\sigma_\perp$ (left) are probed simultaneously.

bulk susceptibility are properly taken into account. If, for example, the first is taken into account by working with spherical samples, but the second disregarded, the NMR rotation patterns can be fitted perfectly, but the results are nonaxially symmetric shielding tensors, which is inconsistent with rapidly reorienting C_5H_5 rings. Consistency is restored by also taking into account the intrinsic anisotropy of the susceptibility.

d. *Protons of Hydrogen Bonds*

These have been the favorite objects of pioneering investigations of proton shielding tensors in solids. With one exception (see below) only hydrogen bonds between oxygen atoms have been studied so far. The salient common features of the results are

(i) The shielding is approximately axially symmetric, although quite often a deviation from axial symmetry is measureable.

(ii) The anisotropy is always large by the standard of isotropic proton shifts. The admittedly still limited data indicate that the anisotropy is larger in ionic systems than in comparable neutral ones. An example of this relation is the intramolecular hydrogen bond in the maleate anion ($\Delta\sigma = \sigma_{zz} - \frac{1}{2}[\sigma_{xx} + \sigma_{yy}] = +30.3$ ppm) and in maleic acid ($\Delta\sigma = +22.6$ ppm).

(iii) The unique shielding direction always lies approximately along the hydrogen bond. As the "direction of a hydrogen bond" is poorly defined our statement actually means approximately along the vector joining the hydrogen bonded atoms.

There are often small but distinct tilts of the unique proton shielding direction to the O···O vector. These tilts carry valuable information about the actual positions of hydrogen atoms in crystal lattices, and about finer details of hydrogen bonds.[128,129]

(iv)   Whenever the hydrogen bond is part of a planar structure such as

the least shielded direction is always found to be close to the normal of the plane.

e. *Example of a Hydrogen Bond between F Atoms: KHF$_2$*

The protons in KHF$_2$ form hydrogen bonds between pairs of fluorine atoms. They have been studied by Van Hecke *et al.*[79] The experiment has many interesting facets that result from the peculiar structure of KHF$_2$. In this substance linear, hydrogen bonded [F···H···F]$^-$ anions are arranged in plane layers. There are two types of crystallographically equivalent, mutually perpendicular anions. The hydrogens are known to occupy symmetric positions between the fluorines. The distance between the hydrogen-bonded fluorine atoms is exceptionally small, $r_{FF} = 2.30$ Å. This makes the dipolar coupling between the bonding hydrogen atom and the bonded fluorines very large. Because of the symmetric position of the proton in the linear [F···H···F]$^-$ anion, the broad-line dipolar spectrum of magnetically equivalent, isolated anions consists of three lines with intensity ratios 1:2:1 arising, respectively, from anions with both fluorine spins up ($\uparrow\uparrow$), down ($\downarrow\downarrow$), and antiparallel ($\uparrow\downarrow, \downarrow\uparrow$).

The separation of the outer lines is

$$\Delta\omega(\Theta) = 2\gamma_H \gamma_F \hbar (3\cos^2\Theta - 1)/r_{HF}^3,$$

where Θ is the angle between the axis of the anion and $\mathbf{B}_0$. For the extreme case $\Theta = 0$, the measured splitting is $2\pi \times 296$ kHz, in excellent agreement with the neutron diffraction value for r_{HF}. Because there are two magnetically inequivalent anions, the dipolar spectrum of a single crystal of KHF$_2$ consists in general of five lines with intensity ratios 1:1:4:1:1. For $\mathbf{B}_0$ parallel to the axis of one type of anions, and hence perpendicular to the other type, all five lines are equally spaced. Such a dipolar spectrum is shown on the top left of Fig. 6-11. For $\mathbf{B}_0$ perpendicular to the layers mentioned above, both types of anions are equivalent and the spectrum consists of three lines (Fig. 6-11, bottom left). What makes KHF$_2$ outstanding with respect to its dipolar spectrum is that for most orientations of $\mathbf{B}_0$ (e.g., those just mentioned) the

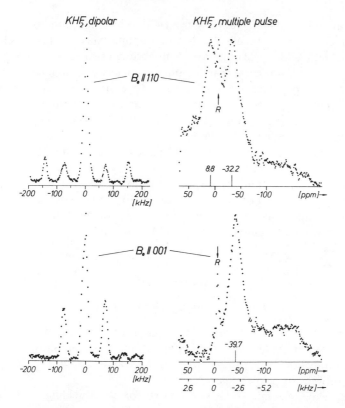

FIG. 6-11. Proton spectra of a single crystal of KHF_2. (*left*) "Wide-line dipolar spectra"; (*right*) proton–fluorine heteronuclear decoupled WAHUHA multiple-pulse spectra (scheme C) with $\tau = 3$ μsec.

(*top*) $\mathbf{B_0}$ oriented so as to be parallel to one type and perpendicular to the other type of $[F \cdots H \cdots F]^-$ anions. (*bottom*) $\mathbf{B_0}$ perpendicular to both types of anions. The positions of the peaks in the multiple-pulse spectra give directly the three principal components of σ_H. The line labeled R arises from a very small amount of water trapped in the crystal. It serves as a convenient internal reference.

line separations are substantially larger than the linewidths. The "broadline" spectra shown in Fig. 6-11 are probably the best resolved dipolar spectra ever observed.

With regard to multiple-pulse spectroscopy and resolution of chemical shifts it should be pointed out, however, that the widths of the lines are rather normal and that it is the separation of lines which is abnormal.

The widths of the lines are due to both ^1H—^{19}F and ^1H—^1H inter-"anionic" coupling. (The proton coupling to the potassium cations is negligible in dipolar spectra, it is not in narrowed multiple-pulse spectra.) Removal of the intra-"anionic" ^{1}H—^{19}F coupling by ^{1}H—^{19}F heterodecoupling in a ^{1}H—^{1}H homonuclear, ^{1}H—^{19}F heteronuclear decoupling experiment is hopeless with currently accessible cycle times for all interesting orientations of $\mathbf{B_0}$.

The very fact, however, that the dipolar lines are so well separated (Fig. 6-11) makes it possible, on the other hand, to do homonuclear–heteronuclear multiple-pulse experiments on just the center line of the dipolar spectrum. This is demonstrated by the multiple-pulse spectra—according to scheme C described in Chapter IV, Section F,5,b—on the right-hand side of Fig. 6-11. Experimental parameters are given in the legend of the figure. The orientations of the crystal were the same as for the dipolar spectra in the same lines of the figure. As crystal symmetry requires the axis of the anion, the in-layer perpendicular to the bond, and the normal to the layer to be principal ^{1}H shielding components, each of the three lines shown gives one of the principal ^{1}H shielding components. Relative to TMS we found

$$\sigma_{xx} = -39.7 \text{ ppm} \quad \text{(perpendicular to the layer)},$$

$$\sigma_{yy} = -32.2 \text{ ppm} \quad \text{(in-layer perpendicular to the bond)},$$

$$\sigma_{zz} = +8.8 \text{ ppm} \quad \text{(along the axis of the anion)}.$$

These numbers imply the largest proton shielding anisotropy found in any substance so far. As far as the orientations of the most, intermediate, and least shielded directions are concerned they fit the rules quoted above for hydrogen bonds between oxygen atoms.

These experimentally established rules for the shielding of protons in hydrogen bonds together with the characteristics of the shielding of aliphatic (methylene group), olefinic, and aromatic protons urgently call for theoretical interpretations. We shall now consider briefly recent ab initio calculations and then turn to less rigorous, but more physical interpretations.

3. THEORETICAL APPROACHES TO PROTON SHIELDING TENSORS

a. *Ab Initio Calculations*

Calculations of proton shielding tensors have been published recently by Ditchfield.[135] His prototype molecules for aliphatic and olefinic protons were—naturally—CH_4 and C_2H_4. For CH_4, where the PAS is fixed by symmetry, he finds $\Delta\sigma = +9.42$ ppm. The experimental results best suited for a comparison are those of succinic acid anhydride ($\Delta\sigma \approx +10$ ppm) and malonic acid (methylene group, $\Delta\sigma \approx +6$ ppm). Theory and experiment agree with respect to the PAS's—which is not surprising because of the importance of (local) symmetry for the shielding—and they also agree with respect to the sign and range of the anisotropy. It is our impression that the

[135] R. Ditchfield, *Chem. Phys.* **2**, 400 (1973).

TABLE 6-4

COMPARISON OF THEORETICAL AND EXPERIMENTAL RESULTS FOR THE SHIELDING OF OLEFINIC PROTONS[a]

	σ_{xx}	σ_{yy}	σ_{zz}	η	
Ethylene	−10	−5.6	−2.38	0.8	theory
Maleic acid	−10.6	−7.9	−4.3	0.82	exp.
Maleate anion	−9	−6.8	−4.1	0.87	exp.
trans-Diiodoethylene	−10.9	−8.4	−6.1	0.93	exp.[b]

[a] ppm relative to TMS.
[b] Preliminary results.

ab initio calculation for CH_4 is better than the comparability of shielding anisotropies of aliphatic protons in different compounds.

Table 6-4 gives a comparison of theoretical and experimental results for olefinic protons. Theory and experiment agree with regard to the least shielded direction. They agree also for olefinic protons in cis-position in that the C=C double bond rather than the C—H bond direction is a principal shielding direction. This is the most important point. Further agreement exists with respect to the asymmetry. All asymmetry parameters η in Table 6-4 are close to unity. The beauty of the overall agreement between theory and experiment is somewhat spoiled by the interchange of the intermediate and least shielded directions. This disagreement should, however, not be overemphasized since the differences of the respective shielding components are very small.

Ethylene

Olefinic fragment
of the maleate anion
and maleic acid

trans-Diiodoethylene

Ditchfield's result for a free H_2O molecule—almost axially symmetric σ tensor, unique direction along the O—H bond to within 5.5°, $\Delta\sigma = +17.8$ ppm—compares favorably with the experimental results on the crystal water molecule of Kieserit,[105] with those for $Ca(OH)_2$,[132] and indeed with those for all protons in hydrogen bonds. In Kieserit the protons of the water molecule are also involved in hydrogen bonds. $Ca(OH)_2$ is the only experimental case so far with —O—H protons not involved in hydrogen bonds.

These comparisons may convey the impression that ab initio calculations can yield reliable shielding anistropy data. However, there is not yet a single compound for a direct comparison between theory and experiment. As we have mentioned already, compounds suited for multiple-pulse work are usually not suited for ab initio calculations, and vice versa.

The major drawback of ab initio calculations as they are usually presented is the lack of physical insight they provide for the reader. Learning that it is the presence of a C=C double bond in C_2H_4 that causes a deshielding in the x direction is of not much help.

Therefore, let us now look into a less rigorous, but more physical approach to proton shielding tensors.

b. *The Atom Multipole Method*[136]

Since the pioneering work of Ramsey[137] nuclear magnetic shielding tensors are usually separated into two terms, $\sigma^{(d)}$ and $\sigma^{(p)}$, which are called, respectively, diamagnetic and paramagnetic. The diamagnetic term has a simple origin: It arises from the Larmor precession of the electrons in the applied magnetic field. It may be expressed in terms of the electronic ground state wavefunction in the absence of B_{st}, $|0\rangle$:

$$\sigma_{xx}^{(d)} = \frac{e^2}{2mc^2} \langle 0| \sum_i \frac{y_i^2 + z_i^2}{r_i^3} |0\rangle, \qquad (6\text{-}1)$$

where e and m are the electronic charge and mass, c is the speed of light, $r_i = (x_i, y_i, z_i)$ the position of electron i, $r_i = |r_i|$.

The Ramsey expression for the paramagnetic term is

$$\sigma_{xx}^{(p)} = \frac{e^2}{2mc^2} \sum_{k>0} \left\{ \frac{\langle 0| (l_{xi}/r_i^3)|k\rangle \langle k| \sum_i l_{xi} |0\rangle}{E_0 - E_k} + \text{c.c.} \right\}, \qquad (6\text{-}2)$$

where k runs over all excited electronic states. This term arises from a coupling of the nuclear spins to the applied field mediated by the orbital angular momenta l_i of the electrons. The nuclear-spin, electron-orbital angular momentum Hamiltonian is

$$\mathscr{H}_{IL} = g\beta\gamma_n \hbar I \cdot \sum_i l_i/r_i^3,$$

where β is the Bohr magneton, and g the electronic g factor. $\mathscr{H}_{IL}$ appears in the first matrix element of $\sigma^{(p)}$. $\mathscr{H}_{LB} = \beta \sum_i l_i \cdot B_{st}$ expresses the coupling of the orbital angular momenta of the electrons to B_{st}. $\mathscr{H}_{LB}$ is recognized in the second matrix element of $\sigma^{(p)}$.

In the absence of an applied field B_{st} the orbital angular momenta l_i are

[136] T. D. Gierke and W. H. Flygare, *J. Amer. Chem. Soc.* **94**, 7277 (1972).
[137] N. F. Ramsey, *Phys. Rev.* **78**, 699 (1950).

totally "quenched" in diamagnetic molecules at rest, or in crystals (see Abragam,[9] Chapter VI). Only the application of an external field $\mathbf{B}_{st}$ introduces a preference quantization axis and "unquenches" orbital electronic angular momenta slightly and proportionally to the strength of $\mathbf{B}_{st}$. This leads to a nuclear spin Hamiltonian term linear in both $\mathbf{I}$ and $\mathbf{B}_{st}$, which means a shielding (or antishielding) term.

Not only $\mathbf{B}_{st}$, but also the molecular angular momentum $\mathbf{J}$ is able to set up a preference quantization axis for the orbital motions of the electrons and can thus lead to slight unquenching. Therefore, it is no wonder that a close relation[136] exists between $\sigma^{(p)}$ and the spin–rotation interaction tensor $\mathbf{C}$ mentioned in Chapter II.

This relation is expressed by

$$\sigma_{xx}^{(p)} = \frac{e^2}{2mc^2} \left[\frac{\Theta_{xx} c}{e\gamma_n \hbar} c_{xx} - \sum_n{}' Z_n \frac{y_n^2 + z_n^2}{r_n^3} \right] \tag{6-3}$$

where Θ_{xx} is the appropriate component of the moment of inertia of the molecule and Z_n the atomic number of the nucleus n. The sum runs over all nuclei of the molecule. $\mathbf{r}_n = (x_n, y_n, z_n)$ specifies the position of the nucleus n. The origin of the coordinates is nucleus A in which we are interested and which is exempt from the summation.

$\mathbf{C}$ is a measurable quantity: It has been measured for a number of light molecules including water[138] and methane.[139] For these molecules we can calculate $\sigma^{(p)}$ as the molecular geometries $(\mathbf{r}_n)$ are known.

For tackling $\sigma^{(d)}$ it is but natural to start with the idea that each of the electrons i "knows" to which nucleus of the molecule or crystal it "belongs" and that its wavefunction is concentrated about that nucleus. Starting from this idea Gierke and Flygare[136] treated $\sigma^{(d)}$ by a moment expansion where the coordinates of the electrons are replaced by the coordinates of the corresponding nuclei. Their expression for $\sigma_{xx}^{(d)}$ of a nucleus A is

$$\sigma_{xx}(A) = \mathrm{I} + \mathrm{II} + \mathrm{III} + \mathrm{IV},$$

where

$$\mathrm{I} = \sigma_{\text{atom}}^{(d)}(A) = e^2/3mc^2 \langle 0| \sum_{i=1}^{Z_A} r_i^{-1} |0\rangle,$$

$$\mathrm{II} = (e^2/2mc^2) \sum_n{}' Z_n r_n^{-3} (y_n^2 + z_n^2),$$

$$\mathrm{III} = (e^2/2mc^2) \sum_n{}' [2r_n^{-3}(y_n \langle y \rangle_n + z_n \langle z \rangle_n) - 3r_n^{-5}(y_n^2 + z_n^2) \mathbf{r}_n \cdot \langle \mathbf{\rho} \rangle_n],$$

$$\mathrm{IV} = (e^2/2mc^2) \sum_n{}' r_n^{-3} \langle \tfrac{1}{3}\rho^2 \rangle_n [2 - 3(y_n^2 + z_n^2)/r_n^2].$$

[138] J. Verhoeven and A. Dymanus, *J. Chem. Phys.* **52**, 3222 (1970).
[139] C. H. Anderson and N. F. Ramsey, *Phys. Rev.* **149**, 14 (1966).

$\langle \mathbf{\rho} \rangle_n$ and $\langle \rho^2 \rangle_n$ are defined below. Terms I–IV have the following physical meaning: I arises from the electrons of nucleus A. The assumption is that the bonded atom A—as all other atoms of the molecule—retains a major share of the free-atom electron distribution. This is usually a good assumption for electron-rich atoms since the inner-shell electrons with their disproportionately large influence on I do not participate in bonds. In some cases of hydrogen atoms it may become questionable.

Terms II–IV originate in the electrons of the other atoms of the molecule. Apart from possibly term I, II is by far the largest diamagnetic term. It corresponds to describing the electron clouds on the nuclei n as point charges $-Z_n \cdot e$ at the sites of the nuclei n. The crucial point is that it is canceled by the second term of $\mathbf{\sigma}^{(p)}$. Indeed, these terms must cancel since nucleus A cannot be affected by equal positive and negative charges at the nuclear sites $\mathbf{r}_n$.

Term III still considers the nth electron cloud as a point charge, but displaced by $\langle \mathbf{\rho} \rangle_n = (\langle x \rangle_n, \langle y \rangle_n, \langle z \rangle_n)$ from $\mathbf{r}_n \cdot Z_n e \langle \mathbf{\rho} \rangle_n$ is the dipole moment of atom n. A table of atom dipole moments can be found in Gierke et al.[140] In all the cases of proton shielding tensors we considered thus far, term III turned out to be but a small correction to term IV, the quadrupole term.

It describes the nth electron cloud as a thin spherical shell of electric charge at a radius $\langle \rho^2 \rangle_n^{1/2}$ from the nth nucleus. Tables of $\langle \rho^2 \rangle_n$ were computed by Malli and Fraga.[141] The symmetry of term IV is such that it does not affect the isotropic shift. However, it turns out, as the following examples will show that it often dominates the anisotropy of the shielding.

Our first example is methane. Except for H_2 it is the only case known to the author for which the proton spin–rotation tensor $\mathbf{C}$ has been measured fully. Note that symmetry restricts the quantities to be measured to $c_{\parallel}$ and $c_{\perp}$. The molecular beam spectroscopy[142] data are [139]

$$c_{\parallel} = (-3.6 \pm 3.2) 2\pi \times 10^3 \, \text{sec}^{-1} \quad \text{and} \quad c_{\perp} = (+17.4 \pm 1.6) 2\pi \times 10^3 \, \text{sec}^{-1},$$

which gives for the first term in Eq. (6-2), the so-called spin–rotation term

$$\sigma_{\parallel}(\text{spin–rotation}) - \sigma_{\perp}(\text{spin–rotation}) = -(22 \pm 5) \, \text{ppm}.$$

As pointed out already, the second term of $\mathbf{\sigma}^{(p)}$ and term II of $\mathbf{\sigma}^{(d)}$ cancel, and the atom dipole term III is negligible. Term IV gives

$$\sigma_{\parallel}^{(d)}(\text{IV}) - \sigma_{\perp}^{(d)}(\text{IV}) = +38 \, \text{ppm},$$

where $\langle \rho^2/3 \rangle_C = 1.2 \times 10^{-16} \, \text{cm}^2$ and $\langle \rho^2/3 \rangle_H = 0.25 \times 10^{-16} \, \text{cm}^2$ have been inserted according to Malli and Fraga.[141]

[140] T. G. Gierke, H. L. Tigelaar, and W. H. Flygare, *J. Amer. Chem. Soc.* **94**, 330 (1972).
[141] G. Malli and S. Fraga, *Theor. Chim. Acta* **5**, 284 (1966).
[142] Note that the molecular beam and the NMR conventions for defining $\mathbf{C}$ differ by a factor 2π.

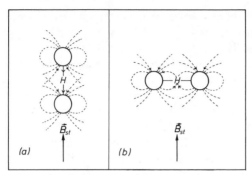

FIG. 6-12. Diamagnetic shielding of protons in hydrogen bonds. (a) $\mathbf{B}_{st}$ parallel to O$\cdots$H$\cdots$O fragment. $\Delta B^{(d)}$ is opposite to $\mathbf{B}_{st}$ at the sites of the oxygen nuclei, and also at the site of the proton. (b) $\mathbf{B}_{st}$ perpendicular to O$\cdots$H$\cdots$O fragment. $\Delta B^{(d)}$ is still opposite to $\mathbf{B}_{st}$ at the site of the oxygen nuclei, but parallel to $\mathbf{B}_{st}$ at the site of the proton.

For the measurable quantity $\Delta\sigma = \sigma_{\|} - \sigma_{\perp}$ we thus get $+(16 \pm 5)$ ppm.

This value is somewhat larger than the result of the ab initio calculation, and also larger than the experimental anisotropies measured for aliphatic protons so far, but this simple analysis at least reproduced correctly the sign of the anisotropy and also roughly its size. We learn from it that the diamagnetic quadrupolar effect for which we have a clear physical comprehension dominates the proton shielding anisotropy in methane.

We next consider protons forming hydrogen bonds between two oxygen atoms. Here, too, we find the shielding anisotropy to be dominated by the quadrupolar term IV of $\sigma^{(d)}$. Due to its $1/r^3$ dependence the major share comes from the neighboring oxygen atoms. For an O$\cdots$H$\cdots$O fragment we can visualize by a classical argument why the bond direction is the most shielded one.

If $\mathbf{B}_{st}$ is along the O—O bond (Fig. 6-12a) the field ΔB induced by the electrons of the oxygen atoms is opposite to $\mathbf{B}_{st}$ at the sites of the oxygen atoms but also along the entire bond. The proton is thus shielded from the external field. If $\mathbf{B}_{st}$ is perpendicular to the O$\cdots$O line (Fig. 6-12b), ΔB is still opposite to $\mathbf{B}_{st}$ at the oxygen nuclei; however, if we now close the magnetic flux lines we see that at the position of the proton ΔB is parallel to $\mathbf{B}_{st}$, which implies an antishielding of the proton. This effect thus tends to make $\sigma_{\|}$ larger than $\sigma_{\perp}$, as observed experimentally.

The proton shielding in an O$\cdots$H$\cdots$O fragment is necessarily axially symmetric. However, the O$\cdots$H$\cdots$O fragment is often part of a planar structure, e.g., in dimers held together by two adjacent hydrogen bonds. Then two principal directions perpendicular to the bond become distinguishable. Our simple magnetostatic picture and, equivalently, term IV of the moment expansion of $\sigma^{(d)}$ predict correctly the direction perpendicular to the plane as the least shielded one.

The perpendicular of planar molecular structures is found to be the least shielded direction not only for protons in hydrogen bonds; according to our current experimental knowledge it is also the least shielded direction for olefinic and aromatic protons. Combining this experimental knowledge with the general behavior of the theoretical term IV leads us to the working hypothesis that the perpendicular of planar molecular structures is generally the least shielded direction for protons located within the plane but at the periphery of the molecule. In this context it is interesting to note that the same direction is typically found to be the most shielded one for ^{13}C and ^{19}F.[14,113,117,143] This contrast between protons on the one hand, and ^{13}C and ^{19}F nuclei on the other hand, can also be accounted for by simple magnetostatic models.[129,144]

We close this chapter with a summary.

D. Summary

1. Spin Dynamics

Multiple-pulse methods have demonstrated that and how Hamiltonians of spin systems can be manipulated by coherent irradiation. This was the central subject of Chapters IV and V. We stress again that the spin Hamiltonians are

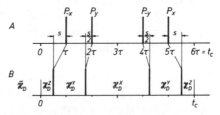

Fig. 6-13. Sequence generating a negative dipolar Hamiltonian. A, WAHUHA sequence, preparation pulse omitted, $\mathscr{H}_D = 0$; B, WAHUHA sequence retimed, $s = \frac{1}{2}\tau$,

$$\bar{\mathscr{H}}_D = \tfrac{1}{6}[\mathscr{H}_D{}^z + \tfrac{5}{2}(\mathscr{H}_D{}^x + \mathscr{H}_D{}^y)] = -\tfrac{1}{4}\mathscr{H}_D{}^z = -\tfrac{1}{4}\mathscr{H}_D.$$

In the limit $s = \tau$ one achieves $\bar{\mathscr{H}}_D = -\tfrac{1}{2}\mathscr{H}_D$. In this limit the $-x$ and $+x$ pulses can be omitted except for the very first $-x$ and very last $+x$ pulses. The P_y^{90} and P_{-y}^{90} can be replaced by P_y^{180} and P_{-y}^{180} bursts lasting for a time 3τ. This last modification has been used by Rhim et al.[40,41]

[143] B. R. Appleman, T. Tokushiro, G. Fraenkel, and C. W. Kern, *J. Chem. Phys.* **60**, 2574 (1974).
[144] H. M. McConnell, *J. Chem. Phys.* **27**, 226 (1957).

manipulated through the spin variables themselves and not only through parameters. In particular, it has been shown that—depending on the particular sequence chosen—certain but different classes of spin–interactions can be suppressed effectively. But this is not all. A minor retiming of the WAHUHA sequence shows (Fig. 6-13) that even the sign of spin–spin interactions can be reversed. For the time evolution of any system the reversal of the sign of the Hamiltonian is equivalent to a reversal of the sign of the time. The possibility of switching the sign of a Hamiltonian led to actual time reversal experiments on systems of many strongly interacting particles,[40,41] which in other fields of physics can only be done as *Gedankenexperimente*. These experiments revealed an intriguing limitation of the fundamental spin temperature hypothesis.[60,145]

2. CHEMICAL ANALYSES

Chemical analyses, which made high-resolution NMR in liquids a big business, have not yet profited from multiple-pulse techniques. This is unfortunate since it keeps the patent-holders still waiting for their royalties.

3. MOLECULAR DYNAMICS

Experiments where information about molecular dynamics has been obtained through the study of multiple-pulse powder spectra have been reviewed in Section B, 4.

4. SHIELDING TENSORS

A substantial number of ^{1}H and ^{19}F shielding tensors have now been measured by multiple-pulse techniques. Proton-enhanced nuclear induction spectroscopy[50] and high-field NMR[37,110] have added a considerable volume of information about ^{13}C, ^{15}N, and ^{31}P shielding tensors. In our judgment it is fair to state that we now have a good general knowledge of the tensorial properties of the magnetic shielding of these nuclei. This means that the first goal of the experimental task of high resolution NMR in solids is completed, or is currently being completed. This achievement should not be underestimated. Nevertheless, the time has now come to ask: What are these shielding tensors good for?

In the introductory parts of many papers from the developmental era of our field, it has been claimed stereotypically that these tensors contain "detailed structural and dynamical" information. Have the experimental results lived up to this expectation?

[145] M. Goldman, "Spin Temperature and Nuclear Magnetic Resonance in Solids." Oxford Univ. Press (Clarendon), London and New York, 1970.

Recalling the results on fluorobenzenes and trifluoroacetic acid, in particular, it is clear that ^{19}F shielding tensors indeed are very sensitive to details of the bonding and packing. A largely unsolved problem, however, is the translation of the structural information contained in these tensors into a more comprehensible form. It is our impression that this problem has prevented the ^{19}F field from booming faster than it actually does.

Regarding protons the experiments and their analyses have revealed in many cases a dominance of the shielding by the simple diamagnetic effect. This is particularly true of protons forming hydrogen bonds. Qualitatively, the shielding of such protons would not be different in hypothetical crystals in which the real atoms are replaced by totally free and therefore spherically symmetric atoms. The chemical bonding of the atoms affects the quantitative features of the proton shielding and we are approaching a state where shift tensors of hydrogen bonds can be related to structural and dynamical details of the bonding, such as whether or not the bonding hydrogen atom is "shared" equally by the bonded oxygen atoms and whether or not it is displaced from the line that geometrically connects the bonded atoms.[106, 128, 129]

In other cases, such as ferrocene, trans-diiodoethylene, and olefinic protons of maleic acid, where the shielding appears not to be dominated by the diamagnetic effect, we are far from even a qualitative understanding of the experimental results. Hopefully this state of affairs will stimulate and challenge the theoretical community, and there will be enough feedback for the multiple-pulse techniques so that these may become firmly established as a useful tool in the study of matter.

Appendix A. Rotations of Angular Momentum Operators

We give here the transformation formulas for rotations about the axes of an orthogonal right-handed frame. Rotations about general axes can be generated by a succession of three rotations about the axes of either a space- or body-fixed frame (cf., e.g., Edmonds[1]):

$$
\exp[i\varphi I_z]
\begin{vmatrix}
I_z \\
I_x \\
I_y \\
I_{+1} \\
I_{-1}
\end{vmatrix}
\exp[-i\varphi I_z] =
\begin{cases}
I_z \\
I_x \cos\varphi - I_y \sin\varphi \\
I_y \cos\varphi + I_x \sin\varphi \\
I_{+1}\exp[i\varphi] \\
I_{-1}\exp[-i\varphi]
\end{cases}
$$

$$
\exp[i\varphi I_x]
\begin{vmatrix}
I_z \\
I_x \\
I_y \\
I_{+1} \\
\\
I_{-1}
\end{vmatrix}
\exp[-i\varphi I_x] =
\begin{cases}
I_z \cos\varphi + I_y \sin\varphi \\
I_x \\
I_y \cos\varphi - I_z \sin\varphi \\
\dfrac{i}{\sqrt{2}} I_z \sin\varphi + I_{+1}\cos^2\varphi/2 - I_{-1}\sin^2\varphi/2 \\
\dfrac{i}{\sqrt{2}} I_z \sin\varphi - I_{+1}\sin^2\varphi/2 + I_{-1}\cos^2\varphi/2
\end{cases}
$$

$$\exp[i\varphi I_y] \begin{vmatrix} I_z \\ I_x \\ I_y \\ I_{+1} \\ \\ I_{-1} \end{vmatrix} \exp[-i\varphi I_y] = \begin{cases} I_z \cos\varphi - I_x \sin\varphi \\ I_x \cos\varphi + I_z \sin\varphi \\ I_y \\ -\dfrac{1}{\sqrt{2}} I_z \sin\varphi + I_{+1}\cos^2\varphi/2 + I_{-1}\sin^2\varphi/2 \\ +\dfrac{1}{\sqrt{2}} I_z \sin\varphi + I_{+1}\sin^2\varphi/2 + I_{-1}\cos^2\varphi/2 \end{cases}$$

Except for the sign in front of $\sin\varphi$ the formulas for I_x, I_y, and I_z are almost trivial and easy to remember.

For the convenience of the reader we also give the conversion formulas for the normalized spherical (I_{+1}, I_0, I_{-1}), unnormalized spherical (I_+, I_-, I_z), and the cartesian components (I_x, I_y, I_z) of $\mathbf{I}$:

$$I_{+1} = -\frac{1}{\sqrt{2}}I_+ = -\frac{1}{\sqrt{2}}(I_x + iI_y),$$

$$I_{-1} = \frac{1}{\sqrt{2}}I_- = \frac{1}{\sqrt{2}}(I_x - iI_y),$$

$$I_0 = I_z,$$

$$I_x = -\frac{1}{\sqrt{2}}(I_{+1} - I_{-1}) = \frac{1}{2}(I_+ + I_-),$$

$$I_y = \frac{i}{\sqrt{2}}(I_{+1} + I_{-1}) = -\frac{i}{2}(I_+ - I_-),$$

$$I_z = I_0.$$

Appendix B. Time Ordering and the Magnus Expansion

We consider the evolution of a system under the influence of its Hamiltonian $\hbar\mathcal{H}$. The system is described by its density matrix $\rho(t)$, which evolves according to the von Neumann equation

$$\dot{\rho}(t) = -i[\mathcal{H}(t), \rho(t)].\tag{B-1}$$

If $\mathcal{H}$ is constant with respect to time, Eq. (B-1) may easily be integrated formally:

$$\rho(t) = U(t)\rho(0)\,U^{-1}(t),\tag{B-2}$$

where

$$U(t) = \exp[-i\mathcal{H}t].\tag{B-3}$$

If $\mathcal{H}$ is not constant in the time interval of interest (say, from $t = 0$ to $t = t_c$), but is constant in smaller intervals $\tau_1, \ldots, \tau_k, \ldots, \tau_n$ with

$$\tau_1 + \cdots + \tau_k + \cdots + \tau_n = t_c,$$

then by repeated integration we obtain

$$U(t_c) = \exp[-i\mathcal{H}_n\tau_n]\cdots\exp[-i\mathcal{H}_k\tau_k]\cdots\exp[-i\mathcal{H}_1\tau_1],\tag{B-4}$$

where we denote by $\mathcal{H}_k$ the Hamiltonian operative in the kth time interval. $U(t_c)$ as given by Eq. (B-4) equals $\exp[-i(\mathcal{H}_n\tau_n + \cdots + \mathcal{H}_k\tau_k + \cdots + \mathcal{H}_1\tau_1)]$ only if all $\mathcal{H}_k$ commute with each other. Finally, if $\mathcal{H}$ varies continuously from $t = 0$ to $t = t_c$, $U(t_c)$ may be expressed formally by

$$U(t_c) = T\exp\left[-i\int_0^{t_c}\mathcal{H}(t)\,dt\right].\tag{B-5}$$

Equation (B-5) is to be considered the limiting case of Eq. (B-4). T is the

Dyson time-ordering operator, which orders operators of greater time arguments in the expanded exponential to the left, e.g.,

$$T\{\mathcal{H}(t')\,\mathcal{H}(t'')\} = \begin{cases} \mathcal{H}(t')\,\mathcal{H}(t'') & \text{if } t' > t'', \\ \mathcal{H}(t'')\,\mathcal{H}(t') & \text{if } t'' > t', \end{cases}$$

but also $T\{\mathcal{H}_2\,\mathcal{H}_1\} = T\{\mathcal{H}_1\,\mathcal{H}_2\} = \mathcal{H}_2\,\mathcal{H}_1$ if $\mathcal{H}_2$ "acts" at later times on the system than does $\mathcal{H}_1$.

Magnus Expansion

The goal of the Magnus expansion is to find an expression in the form of $\exp[-iFt_c]$ for $U(t_c)$ as given by Eq. (B-4) or (B-5). In order to find F we start by expanding Eq. (B-4) in a power series:

$$U(t_c) = [\cdots(n)\cdots] \times \cdots \times [\cdots(k)\cdots] \times \cdots$$

$$\times \left[1 + (-i\mathcal{H}_1\tau_1) + \frac{(-i_1\mathcal{H}_1\tau_1)^2}{2!} + \frac{(-i_1\mathcal{H}_1\tau_1)^3}{3!} + \cdots \right], \quad \text{(B-6)}$$

where (k) means that the indices of $\mathcal{H}$ and τ are k.

We consider $\mathcal{H}_k\tau_k$ as a small quantity of first order. In case it is not, we simply subdivide the time interval $0, \ldots, t_c$ further until $\mathcal{H}_k\tau_k$ is small. By rearranging Eq. (B-6) we get

$$U(t_c) = 1$$

$$+ (-i)\{\mathcal{H}_1\tau_1 + \mathcal{H}_2\tau_2 + \cdots + \mathcal{H}_n\tau_n\}$$

$$+ \frac{(-i)^2}{2!}(\mathcal{H}_1{}^2\tau_1{}^2 + \cdots + \mathcal{H}_n{}^2\tau_n{}^2 + 2[\mathcal{H}_2\mathcal{H}_1\tau_2\tau_1 + \cdots + \mathcal{H}_n\mathcal{H}_{n-1}\tau_n\tau_{n-1}])$$

$$+ \frac{(-i)^3}{3!}(\mathcal{H}_1{}^3\tau_1{}^3 + \cdots + 3[\mathcal{H}_2\mathcal{H}_1{}^2\tau_2\tau_1{}^2 + \mathcal{H}_2{}^2\mathcal{H}_1\tau_2{}^2\tau_1 + \cdots]$$

$$+ 6[\mathcal{H}_3\mathcal{H}_2\mathcal{H}_1\tau_3\tau_2\tau_1 + \cdots])$$

$$+ \cdots. \quad \text{(B-7a)}$$

Note that in all products the operators are ordered in ascending order to the left! Equation (B-7a) can be rewritten in a compact form as follows:

$$U(t_c) = 1 + (-i)\{\cdots\} + \frac{(-i)^2}{2!}T\{\cdots\}^2 + \cdots + \frac{(-i)^m}{m!}T\{\cdots\}^m + \cdots, \quad \text{(B-7b)}$$

where $\{\cdots\} = \{\mathcal{H}_1\tau_1 + \mathcal{H}_2\tau_2 + \cdots + \mathcal{H}_n\tau_n\}$.

For example, $T\{\cdots\}^2 = \cdots T(\mathcal{H}_1 \mathcal{H}_2 + \mathcal{H}_2 \mathcal{H}_1)\tau_2\tau_1 = \cdots 2\mathcal{H}_2\mathcal{H}_1\tau_2\tau_1$ as it should according to Eq. (B-7a).

Next we make an ansatz for F:

$$F = \bar{\mathcal{H}} + \bar{\mathcal{H}}^{(1)} + \bar{\mathcal{H}}^{(2)} + \cdots. \tag{B-8}$$

In this ansatz we assume $t_c\bar{\mathcal{H}},\ldots,t_c\bar{\mathcal{H}}^{(k)},\ldots$ to be small quantities of first, $\ldots,k$th, $\ldots$ order. We shall confirm this assumption later. Now, as above, we expand $\exp[-iFt_c]$:

$$\exp[-iFt_c] = \exp[-it_c(\bar{\mathcal{H}} + \bar{\mathcal{H}}^{(1)} + \bar{\mathcal{H}}^{(2)} + \cdots)]$$

$$= 1$$

$$+ (-i)t_c\left\{\underset{①}{\bar{\mathcal{H}}} + \underset{②}{\bar{\mathcal{H}}^{(1)}} + \underset{③}{\bar{\mathcal{H}}^{(2)}} + \underset{④}{\bar{\mathcal{H}}^{(3)}} + \cdots\right\}$$

$$+ \frac{(-it_c)^2}{2!}\left\{\underset{②}{(\bar{\mathcal{H}})^2} + \underset{③}{\bar{\mathcal{H}}\bar{\mathcal{H}}^{(1)}} + \underset{③}{\bar{\mathcal{H}}^{(1)}\bar{\mathcal{H}}} + \underset{④}{(\bar{\mathcal{H}}^{(1)})^2}\right.$$

$$\left. + \underset{④}{\bar{\mathcal{H}}\bar{\mathcal{H}}^{(2)}} + \underset{④}{\bar{\mathcal{H}}^{(2)}\bar{\mathcal{H}}} + \underset{⑤}{\bar{\mathcal{H}}^{(2)}\bar{\mathcal{H}}^{(1)}} + \cdots\right\}$$

$$+ \frac{(-it_c)^3}{3!}\left\{\underset{③}{(\bar{\mathcal{H}})^3} + \underset{④}{(\bar{\mathcal{H}})^2\bar{\mathcal{H}}^{(1)}} + \underset{④}{\bar{\mathcal{H}}^{(1)}(\bar{\mathcal{H}})^2}\right.$$

$$\left. + \underset{④}{\bar{\mathcal{H}}\bar{\mathcal{H}}^{(1)}\bar{\mathcal{H}}} + \underset{⑥}{(\bar{\mathcal{H}}^{(1)})^3} + \cdots\right\}$$

$$+ \frac{(-it_c)^4}{4!}\left\{\underset{④}{(\bar{\mathcal{H}})^4} + \cdots\right\}$$

$$+ \cdots. \tag{B-9}$$

The encircled numbers indicate of which order the corresponding terms are small.

We rearrange terms in descending order of magnitude:

$$\exp[-iFt_c] = 1 \qquad\qquad\qquad ⓪$$

$$- it_c\bar{\mathcal{H}} \qquad\qquad\qquad ①$$

$$- it_c\bar{\mathcal{H}}^{(1)} + \frac{(-it_c)^2}{2!}(\bar{\mathcal{H}})^2 \qquad\qquad ②$$

$$- it_c\bar{\mathcal{H}}^{(2)} + \frac{(-it_c)^2}{2!}(\bar{\mathcal{H}}\bar{\mathcal{H}}^{(1)} + \bar{\mathcal{H}}^{(1)}\bar{\mathcal{H}}) + \frac{(-it_c)^3}{3!}(\bar{\mathcal{H}})^3, \qquad ③$$

etc.

$$\tag{B-10}$$

By equating terms of equal order in Eqs. (B-10) and (B-7a) [or (B-7b)] we find

$$\bar{\mathcal{H}} = \frac{1}{t_c}\{\cdots\} = \frac{1}{t_c}\{\mathcal{H}_1\tau_1 + \mathcal{H}_2\tau_2 + \cdots + \mathcal{H}_n\tau_n\}, \tag{B-11}$$

$$\bar{\mathcal{H}}^{(1)} = \frac{-i}{2t_c}[T\{\cdots\}^2 - (t_c\,\bar{\mathcal{H}})^2] = \frac{-i}{2t_c}[T\{\cdots\}^2 - \{\cdots\}^2] \tag{B-12a,b}$$

$$= -\frac{i}{2t_c}([\mathcal{H}_2,\mathcal{H}_1]\tau_2\tau_1 + [\mathcal{H}_3,\mathcal{H}_1]\tau_3\tau_1 + [\mathcal{H}_3,\mathcal{H}_2]\tau_3\tau_2 + \cdots), \tag{B-12c}$$

$$\bar{\mathcal{H}}^{(2)} = -\frac{1}{6t_c}(T\{\cdots\}^3 - (t_c\,\bar{\mathcal{H}})^3 - 3it_c^{\,2}[\bar{\mathcal{H}}\bar{\mathcal{H}}^{(1)} + \bar{\mathcal{H}}^{(1)}\bar{\mathcal{H}}]) \tag{B-13a}$$

$$= -\frac{1}{6t_c}\left(T\{\cdots\}^3 + 2\{\cdots\}^3 - \frac{3}{2}[\{\cdots\}T\{\cdots\}^2 + T\{\cdots\}^2\{\cdots\}]\right) \tag{B-13b}$$

$$= -\frac{1}{6t_c}(\{[\mathcal{H}_3,[\mathcal{H}_2,\mathcal{H}_1]] + [[\mathcal{H}_3,\mathcal{H}_2],\mathcal{H}_1]\}\tau_3\tau_2\tau_1 + \cdots$$

$$+ \frac{1}{2}\{[\mathcal{H}_2,[\mathcal{H}_2,\mathcal{H}_1]]\tau_2^{\,2}\tau_1 + [[\mathcal{H}_2,\mathcal{H}_1],\mathcal{H}_1]\tau_2\tau_1^{\,2} + \cdots\}), \tag{B-13c}$$

etc. From Eqs. (B-12) and (B-13) it is clear that $t_c\,\bar{\mathcal{H}}^{(1)}$ and $t_c\,\bar{\mathcal{H}}^{(2)}$ are small quantities of second and third order, if $t_c\,\bar{\mathcal{H}}$ is a small quantity of first order— as we actually assumed—and so our ansatz is consistent.

If $\mathcal{H}(t)$ jumps only once in the interval $0,\ldots,t_c$, and $\mathcal{H}_1\tau_1 = A$, $\mathcal{H}_2\tau_2 = B$, we find

$$t_c\,\bar{\mathcal{H}} = A + B, \qquad t_c\,\bar{\mathcal{H}}^{(1)} = -\frac{i}{2}[B,A],$$

and

$$t_c\,\bar{\mathcal{H}}^{(2)} = -\tfrac{1}{12}\{[B,[B,A]] + [[B,A],A]\}.$$

It follows that

$$e^{-iB}e^{-iA} = \exp\left(-i(A+B) - \frac{1}{2}[B,A] + \frac{i}{12}\{[B,[B,A]] + [[B,A],A]\} + \cdots\right) \tag{B-14}$$

Equation (B-14) is the Baker–Campbell–Hausdorff formula.

To arrive at the Magnus formula we must subdivide the interval $0,\ldots,t_c$ more and more finely. The limiting case of Eq. (B-11) for all $\tau_k \to 0$ is

$$\bar{\mathcal{H}} = \frac{1}{t_c}\int_0^{t_c} \mathcal{H}(t)\,dt. \tag{B-15}$$

Thus $\bar{\mathscr{H}}$ ("$\mathscr{H}$-bar") is the average of $\mathscr{H}(t)$ in the interval $0, \ldots, t_c$ and is therefore called the average Hamiltonian.

Similarly, the limiting case of Eq. (B-12) is

$$\bar{\mathscr{H}}^{(1)} = \frac{-i}{2t_c} \int_0^{t_c} dt_2 \int_0^{t_2} dt_1 \, [\mathscr{H}(t_2), \mathscr{H}(t_1)]. \tag{B-16}$$

The domain of integration (see Fig. B-1) ensures that everywhere the operator with the greater time argument is on the left—where it should be.

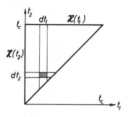

FIG. B-1. Domain of integration for $\bar{\mathscr{H}}^{(1)}$.

The limiting case of Eq. (B-13) is

$$\bar{\mathscr{H}}^{(2)} = -\frac{1}{6t_c} \int_0^{t_c} dt_3 \int_0^{t_3} dt_2 \int_0^{t_2} dt_1$$

$$\times \, \{[\mathscr{H}(t_3), [\mathscr{H}(t_2), \mathscr{H}(t_1)]] + [[\mathscr{H}(t_3), \mathscr{H}(t_2)], \mathscr{H}(t_1)]\}. \tag{B-17}$$

The domain of integration (see Fig. B-2) ensures again that everywhere the operators with greater time arguments are ordered to the left. It also takes care of the factor $\frac{1}{2}$ in front of the second line of Eq. (B-13c).

For the reader who is not satisfied with our claim that Eqs. (B-15)–(B-17) are the limiting cases of Eqs. (B-11)–(B-13) we recommend that he try to recover the latter from the former in a case where $\mathscr{H}(t)$ jumps n times ($n = 5$, say) in the interval considered.

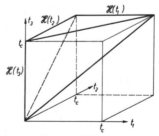

FIG. B-2. Domain of integration for $\bar{\mathscr{H}}^{(2)}$. (Rear, upper left corner of cube).

We proceed to indicate proofs for the special cases mentioned in Chapter IV, Section D. By comparing Eqs. (B-10) and (B-7b) for $\mathscr{H}^{(k)} = 0$, $k < n$, the validity of

$$t_c \bar{\mathscr{H}}^{(n)} = \frac{(-i)^n}{(n+1)!} T\{\cdots\}^{n+1} \qquad (\mathscr{H}^{(k)} = 0 \quad \text{for} \quad k < n) \qquad \text{(B-18)}$$

becomes immediately evident.

As

$$\frac{(-i)^n}{(n+1)!} T\{\cdots\}^{n+1} \xrightarrow[\text{all } \tau_k \to 0]{} (-i)^n \int_0^{t_c} dt_{n+1} \int_0^{t_{n+1}} dt_n \cdots \int_0^{t_2} dt_1$$
$$\times \mathscr{H}(t_{n+1}) \mathscr{H}(t_n) \cdots \mathscr{H}(t_2) \mathscr{H}(t_1), \qquad \text{(B-19)}$$

we establish Eq. (4-44). Note there is no denominator $(n+1)!$ in the right-hand expression of Eq. (B-19). However, the volume of integration is "only" $t_c^{n+1}/(n+1)!$.

Before we turn to the special properties of symmetric cycles we prove a lemma. Consider $\exp[i\mathscr{H}_n \tau_n] \cdots \exp[i\mathscr{H}_2 \tau_2] \exp[i\mathscr{H}_1 \tau_1]$. We call this expression $U(-t_c)$ but avoid assigning any physical meaning to it. For $U(-t_c)$ we make the ansatz $\exp[ift_c]$, and for f we make the ansatz

$$f = \bar{h} + \bar{h}^{(1)} + \bar{h}^{(2)} + \cdots. \qquad \text{(B-20)}$$

The lemma we want to prove is

Lemma 1:

$$\bar{h}^{(k)} = (-1)^k \bar{\mathscr{H}}^{(k)}.$$

For the sake of compactness of notation we define $\bar{\mathscr{H}}^{(0)} = \bar{\mathscr{H}}$. Furthermore let us define the symbol $\mathscr{H}_n^{k+1}$ as the sum of all products of the type $\mathscr{H}^{(k_1)} \mathscr{H}^{(k_2)} \cdots \mathscr{H}^{(k_n)}$ in Eq. (B-9) that contain n factors and that are small of order $m = k+1$. Examples are

$$\mathscr{H}_3^3 = (\mathscr{H}^{(0)})^3, \qquad \mathscr{H}_2^3 = \mathscr{H}^{(0)} \mathscr{H}^{(1)} + \mathscr{H}^{(1)} \mathscr{H}^{(0)}, \qquad \mathscr{H}_1^3 = \mathscr{H}^{(2)}.$$

Correspondingly we define the symbol h_n^{k+1}; it refers, of course, to an equation analogous to Eq. (B-9) in which the signs of all the i's are reversed.

As

$$n + \sum_i k_i = (k_1+1) + (k_2+1) + \cdots + (k_n+1) = m = k+1, \qquad \text{(B-21)}$$

it follows from Lemma 1—if it is correct—that

$$h_n^{k+1} = (-1)^{k_1+k_2+\cdots+k_n} \mathscr{H}_n^{k+1} = (-1)^{k+1-n} \mathscr{H}_n^{k+1}$$

or

$$-(-1)^n \mathscr{H}_n^{k+1} = (-1)^k h_n^{k+1}. \qquad \text{(B-22)}$$

To prove Lemma 1 consider the construction laws of $\bar{\mathscr{H}}^{(k)}$ and $\bar{h}^{(k)}$ [cf. Eqs. (B-9) and (B-7b), and the corresponding equations with the signs of all i's reversed]:

$$\bar{\mathscr{H}}^{(k)} = \frac{(-i)^k}{(k+1)!}\frac{1}{t_c}T\{\cdots\}^{k+1} - \frac{(-it_c)^k}{(k+1)!}\mathscr{H}_{k+1}^{k+1} - \cdots$$
$$- \frac{(-it_c)^2}{3!}\mathscr{H}_3^{k+1} - \frac{(-it_c)}{2!}\mathscr{H}_2^{k+1}, \tag{B-23}$$

$$\bar{h}^{(k)} = \frac{(i)^k}{(k+1)!}\frac{1}{t_c}T\{\cdots\}^{k+1} - \frac{(it_c)^k}{(k+1)!}h_{k+1}^{k+1} - \cdots$$
$$- \frac{(it_c)^2}{3!}h_3^{k+1} - \frac{it_c}{2!}h_2^{k+1}. \tag{B-24}$$

By taking into account Eq. (B-22) we recognize that $\bar{\mathscr{H}}^{(k)}$ and $\bar{h}^{(k)}$ differ, term by term, by $(-1)^k$. Therefore, we recover Lemma 1 for k, having made use of it only for $k' < k$. Recall that the lowest value of n in Eqs. (B-23) and (B-24) is two, and that the largest value of k_i occurring in $\mathscr{H}_2^{k+1}$ and h_2^{k+1} is $k_{max} = k-1$ [see Eq. (B-21)]. As Lemma 1 holds obviously for $k = 0$ ($\bar{\mathscr{H}} = \bar{h}$) we conclude, by induction, that it holds for all k.

Now consider symmetric cycles,[146] i.e., cycles for which $\mathscr{H}(t) = \mathscr{H}(t_c-t)$. An obvious consequence of the symmetry of the cycles is

$$U^{-1}(t_c) = \exp[i\mathscr{H}_1\tau_1] \cdots \exp[i\mathscr{H}_n\tau_n]$$

$\qquad = \exp[iF_s t_c]$ $\qquad\qquad$ [consequence of Eq. (B-4)]

$\qquad = \exp[i\mathscr{H}_n\tau_n] \cdots \exp[i\mathscr{H}_1\tau_1]$ $\quad$ (because the cycle is symmetric)

$\qquad = U(-t_c)$ $\qquad\qquad\qquad$ (by definition)

$\qquad = \exp[if_s t_c]$ $\qquad\qquad\quad$ (by definition).

The indices s on F_s and f_s stand for "symmetric."
We conclude $F_s = f_s$, or

$$\sum_{k=0}^{\infty} \bar{\mathscr{H}}^{(k)} = \sum_{k=0}^{\infty} \bar{h}^{(k)},$$

and with the aid of Lemma 1

$$\bar{\mathscr{H}}^{(k)} = (-1)^k \bar{\mathscr{H}}^{(k)},$$

which requires

$$\bar{\mathscr{H}}^{(k)} = 0 \qquad \text{for} \quad k \text{ odd}.$$

---

[146] From now on we follow closely Wang and Ramshaw.[66]

For the sake of completeness we also consider briefly antisymmetric cycles, i.e., cycles for which $\mathscr{H}(t) = -\mathscr{H}(t_c - t)$.

$$U(t_c) = \exp[-i\mathscr{H}_n \tau_n] \exp[-i\mathscr{H}_{n-1} \tau_{n-1}] \cdots \exp[-i\mathscr{H}_2 \tau_2] \exp[-i\mathscr{H}_1 \tau_1]$$

$$= \exp[i\mathscr{H}_1 \tau_1] \exp[i\mathscr{H}_2 \tau_2] \cdots \exp[i\mathscr{H}_{n-1} \tau_{n-1}] \exp[i\mathscr{H}_n \tau_n]$$
$$\text{(consequence of antisymmetry)}$$

$$= U^{-1}(t_c),$$

or

$$\exp[-iF_a t_c] = \exp[+iF_a t_c]$$

(a stands for antisymmetric). It follows that

$$F_a = -F_a = 0,$$

which means that for antisymmetric cycles $\mathscr{H}$ vanishes together with all its correction terms. Whereas symmetric cycles are of great practical importance in high-resolution NMR in solids, antisymmetric cycles are not, for obvious reasons.

Appendix C. Off-Resonance Averaging of the Second-Order Dipolar Hamiltonian

In this appendix we indicate a proof for the elimination of the on-resonance second-order dipolar Hamiltonian $\mathcal{H}_D^{(2)}$ in a WAHUHA experiment for a continuous rotation about the rotating frame 111 axis. Such a rotation—not continuous, but stepwise—is obtained by going off resonance.

On-resonance $\mathcal{H}_D^{(2)}$ is proportional to

$$[(\mathcal{H}_D^x - \mathcal{H}_D^z), [\mathcal{H}_D^x, \mathcal{H}_D^y]]$$

$$\propto b^{ik}[(I_x{}^i I_x{}^k - I_z{}^i I_z{}^k), [b^{mn}(\mathbf{I}^m \cdot \mathbf{I}^n - 3I_x{}^m I_x{}^n), b^{pq}(\mathbf{I}^p \cdot \mathbf{I}^q - 3I_y{}^p I_y{}^q)]]$$

[cf. Eq. (5-1)]. We start by expressing $\mathcal{H}_D^{(2)}$ by spin operators I_X, I_Y, I_Z referenced to the magic-angle tilted rotating frame. The transformation formulas are given in Chapter IV, Section C,2,d. The double commutator can then be expressed by

$$[(D+C), [(D+C/3), (D-C/3)]] \propto [(D+C), [C,D]],$$

where

$$D = b^{ik}\{I_X{}^i I_X{}^k - I_Y{}^i I_Y{}^k - \sqrt{2}(I_X{}^i I_Z{}^k)_+\}$$

$$C = b^{ik}\{(I_X{}^i I_Y{}^k)_+ + \sqrt{2}(I_Y{}^i I_Z{}^k)_+\}\frac{1}{\sqrt{3}}.$$

Here $(I_X{}^i I_Z{}^k)_+$ is an abbreviation for $(I_X{}^i I_Z{}^k + I_Z{}^i I_X{}^k)$.

The effect of a rotation with angular velocity ω about the rotating frame 111, or magic-angle tilted rotating frame Z-axis is

$$I_X{}^i \rightarrow I_X{}^i \cos \omega t + I_Y{}^i \sin \omega t, \qquad I_Y{}^i \rightarrow I_Y{}^i \cos \omega t - I_X{}^i \sin \omega t, \qquad I_Z{}^i \rightarrow I_Z{}^i.$$

Inserting these expressions into $[C, D]$ results in terms with three different types of time dependences, we have

    (i)   constant in time,

    (ii) $(\cos \omega t \sin 2\omega t + \sin \omega t \cos 2\omega t)$,

    (iii) $(\sin \omega t \sin 2\omega t - \cos \omega t \cos 2\omega t)$.

$D + C$ has terms with time dependences $\sin \omega t$, $\cos \omega t$, $\sin 2\omega t$, and $\cos 2\omega t$. The time average of all products of one of expressions (i)–(iii) with $\sin \omega t, \ldots$, $\cos 2\omega t$ vanishes; hence we conclude that the on-resonance second-order dipolar Hamiltonian is eliminated by going off resonance.

Appendix D. Phase Transients

If one routes an rf pulse through a phase-sensitive detector (PSD) one naively expects at its output a dc pulse with a shape equal to the envelope of the rf pulse. If one varies the phase φ of the local oscillator of the PSD one expects that the shape of the dc pulse will remain unchanged but that its size will vary as $\cos(\varphi - \varphi_0)$, where φ_0 is the phase of the rf carrier of the rf pulse. In particular, one expects that there will be some phase φ which nullifies the dc signal.

However, this is not what one typically observes in practice and the trouble is usually not connected with the PSD. Figure D-1 shows what one typically gets instead. Both cases (a) and (b) are encountered in real life.

The interpretation of Fig. D-1 is that the rf pulse contains out-of-phase (quadrature) components at both its leading and trailing edges. These are called phase transients or phase glitches.

Whatever the nature of their origin is, their effect upon the nuclear magnetization $\langle \mathbf{M} \rangle$ is the following: A pulse that ideally rotates $\langle \mathbf{M} \rangle$ in the rotating frame through an angle β_x about the x axis actually first rotates

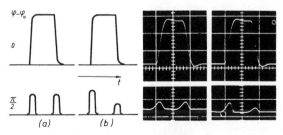

FIG. D-1. Phase transients. Phase-sensitively detected rf pulses are shown with the reference phase φ chosen such that the amplitude of the stationary part of the pulse is at its maximum (*top*) and zero (*bottom*). (a) Symmetric, (b) nonsymmetric phase transients. The right-hand part of the figure shows actual oscilloscope traces.

$\langle \mathbf{M} \rangle$ through a (small) angle α_l about the y axis, then through an angle β_x about the x axis, and finally through a (small) angle α_{tr} about the y axis.

The indices l and tr stand for leading and trailing. We cannot predict the signs of α_l and α_{tr}. From the areas under the bottom traces of the right-hand part of Fig. D-1 one may estimate the size of α_l and α_{tr} to be roughly 3°. (The areas under the top traces correspond to $\beta_x \approx 90°$.) Note, however, that this number is typical for our spectrometer; it may well be appreciably different for others.

To understand how phase transients affect pulse cycles, let us consider the simple phase-alternated $[P_x^{90} - \tau - P_{-x}^{90} - \tau -]_n$ pulse sequence sketched in Fig. D-2.

We suppose that the phase transients are equal with respect to the stationary parts of the pulses for both types of pulses. Further below, we comment upon the justification of this supposition. For case (a) where $\alpha_l = \alpha_{tr}$ we immediately deduce from Fig. D-2 that rotations ③ and ④, ② and ⑤, ① and ⑥ each cancel and that the net rotation of the nuclear magnetization is zero. Note that for rotations ① and ⑥ to cancel there are two conditions. (i) Rotations ① and ⑥, taken by themselves, must cancel and (ii) the series of rotations ②–⑤ between rotations ① and ⑥ must cancel also. Both conditions are fulfilled for symmetric phase transients and correctly adjusted 90° pulses. We conclude that symmetric phase transients ($\alpha_l = \alpha_{tr}$) do not destroy the cyclic property of the phase-alternated pulse sequence. This conclusion applies also for more complex multiple-pulse sequences such as the WAHUHA and MREV sequences: Symmetric phase transients are not harmful.

We now turn to case (b) where $\alpha_l \neq \alpha_{tr}$. The adjacent rotations ③ and ④ do not cancel. Rotations ① and ④, which are about the same axis and which carry $\langle \mathbf{M} \rangle$ through equal angles in opposite directions, would cancel if they were not separated by another rotation about a different axis, rotation ②. The trouble is, of course, that rotations about different axes simply do not

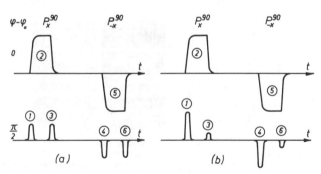

FIG. D-2. Phase-alternated pulse sequence $[P_x^{90} - \tau - P_{-x}^{90} - \tau]_n$. (a) Symmetric, (b) non-symmetric phase transients.

commute. Thus, the net rotation of $\langle \mathbf{M} \rangle$ is nonzero in case (b) and we conclude that nonsymmetric phase transients ($\alpha_l \neq \alpha_{tr}$) destroy the cyclic property of the phase-alternated pulse sequence, and the same holds true of more complex multiple-pulse sequences. Nonsymmetric phase transients are harmful.

We recognized the harmful effects of nonsymmetric phase transients in the early days of multiple-pulse sequences and described tricks to compensate for them.[147] These tricks are somewhat obsolete now since means have been found (see below) that enable us to avoid nonsymmetric phase transients from the very beginning. To understand them, a knowledge of the origin of phase transients is required.

Ellett[92] has shown that phase transients are inevitable companions of rf pulses in resonance circuits.[148] He points out, however, that by properly tuning the resonance circuits the phase transients can be made symmetric ($\alpha_l = \alpha_{tr}$) and thus harmless.

Mehring and Waugh[149] have considered rf switching—as done in every coherent pulse spectrometer—as a source of phase transients. Double-balanced mixers are now used almost universally for rf switching. Their switching speed is very high, and the transition from "off" to "on" and from "on" to "off" is typically shorter than the period of the rf carrier. Hence we feel that this effect can hardly account for the phase transients observed finally in the NMR probe, where they last for many periods of the rf carrier.

In our pulse spectrometer, at least, the major producers of phase transients are broad-band, but tuned, class C transistor amplifiers driven into saturation. Empirically we discovered the ratio of the leading and trailing phase transients (α_l/α_{tr}) to depend very sensitively on

(a) the level of the driving rf power, and
(b) the quiescent base currents of the transistors.

These transistor amplifiers are used at power levels of ≈ 50 mW to 10 W and the phase transients they produce propagate right through the following tube power amplifiers (2.5 kW) into the NMR probe.

By monitoring the quiescent base currents of the relevant transistors we are able to vary the ratio α_l/α_{tr} of the phase transients in the NMR probe smoothly from well below to well above unity without affecting the rf-pulse power virtually at all.

Of course, what is called for is to have the phase transients symmetric and thus harmless simultaneously for all the $x, -x, y,$ and $-y$ pulses. By monitoring the quiescent base currents of the relevant transistors as described, we can

[147] J. S. Waugh and U. Haeberlen, U.S. Patent 3,530,374 (1970).
[148] He thinks—and so do we—that professional electronics engineers were aware of that fact long ago.
[149] M. Mehring and J. S. Waugh, *Rev. Sci. Instrum.* **43**, 649 (1972).

adjust for this condition only if the levels of all pulses are the same at the inputs of the transistor amplifiers in question. It is by no means a matter of course that they are the same, since in all multiple-pulse spectrometers the x, $-x$, y, and $-y$ pulses are routed through different channels with different gains before they are combined and further amplified in a single-channel power amplifier chain.

Thus we see that phase transients are one reason—but only one—why multiple-pulse spectrometers must be designed very carefully even in parts that are absolutely uncritical for more standard types of pulsed NMR experiments.

Subject Index

A

Ab initio calculations, of shielding tensors, 162–164
Angular momentum operators, rotations of, 171–172
Antisymmetric tensor constituents, 8, 31–35
Aromatic protons, 158–159
Atom multipole method, 164–168
Average Hamiltonian approximation of WAHUHA sequence, 53–58, 93
Average Hamiltonian theory, 47, 64–69
   cyclic interactions in, 65–67
   Magnus expansion in, 67–69
   symmetric and antisymmetric cycles in, 69
Average pulse error Hamiltonian,
   for MREV sequence, 119
   for WAHUHA sequence, 117
Averaging
   coherent, 4, 72–74
   by random anisotropic motions, 39–41
   by random isotropic motions, 39
   by sample spinning, 42–47
   in spin space, 47–64, 74–89
   stochastic, 69–74

B

Bloch–Siegert shift, 77–82

C

Calcium fluoride, 128–130
Carr–Purcell sequence, 74–77
Chemical exchange, 41
Coherent averaging vs. stochastic averaging, 69–74
CP sequence, see Carr–Purcell sequence
Cross terms, in multiple-pulse experiments
   dipolar-resonance offset, 123–124
   dipolar-shielding, 95–96
CRS (crystal fixed orthogonal axis system), 20

Crystallographical equivalence, 17–19
Cycle, 66
   antisymmetric, symmetric, 69, 179–180
Cyclic interactions, 65–67
Cyclohexane, 42

D

Decoupling
   heteronuclear, 82–87
   homonuclear, 3, 82
   self, 87–89
Dipolar Hamiltonian $\hbar\mathcal{H}_{CD}$, 7, 11–14
Dipolar line broadening, residual, 112
Dipolar linewidth, 137–139
Double quantum transitions, 85

E

Equivalence
   crystallographical, 17–18
   magnetical, 17–18
Evan's eight-pulse cycle, 134

F

Ferrocene, 129, 158–159
FID, see Free induction decay
Finite pulse widths, 74–76, 105–110
Flip angle, 55
Flip angle errors, 76–77, 110–112, 114–125
   compensation of, 119–120
Fluorine-19, in multiple pulse techniques, 138–149
Fourier transform NMR experiment, 53–57
Free induction decay, 53, 88–89
Frequency shifts, from cw rf fields, 77–82

G

Gill–Meiboom modification of CP sequence, 76–77